投考公務員
題解 EASY PASS
英文運用

Ray Leung 著

U0130780

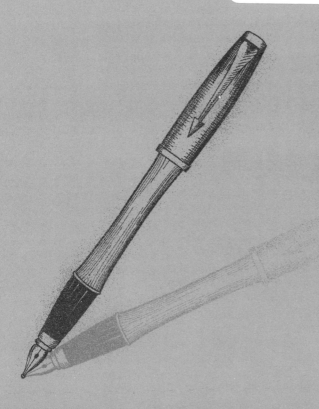

CONTENT 目錄

PART I CRE 基礎概念

公務員綜合招聘考試須知

綜合招聘考試(Common Recruitment Examination, 簡稱CRE)包括三張各為45分鐘的多項選擇題試卷，分別是：

1) 英文運用

2) 中文運用

3) 能力傾向測試

目的是評核考生的英、中語文能力及推理能力。英文運用及中文運用試卷的成績分為二級、一級或不及格，並以二級為最高等級；而能力傾向測試的成績則分為及格或不及格。英文運用及中文運用試卷的二級及一級成績和能力傾向測試的及格成績永久有效。

基本法及香港國安法測試

基本法及香港國安法是一張設有中英文版本的選擇題形式試卷。全卷共20題，考生須於30分鐘內完成。考生如在20題中答對10題或以上，會被視為取得《基本法及香港國安法》測試的及格成績。

CRE形式

試卷的試題類型及題目數量如下：

試卷 (多項選擇題)	題目數量	時間	試題類型
中文運用	45	45分鐘	閱讀理解 字詞辨識 句子辨析 詞句運用
英文運用 Use of English	40	45分鐘	Comprehension Error Identification Sentence Completion Paragraph Improvement
能力傾向測試	35	45分鐘	演繹推理 · Verbal Reasoning(English) · Numerical Reasoning · Data Sufficiency Test · Interpretation of Tables and Graphs

公開試成績與轄免

香港中學文憑考試英國語文科第5級或以上成績，會獲接納為等同綜合招聘考試英文運用試卷的二級成績。香港中學文憑考試中國語文科第5級或以上成績會獲接納為等同綜合招聘考試中文運用試卷的二級成績。持有上述成績的申請人,將不會被安排應考英文運用及/ 或中文運用試卷。

香港高級程度會考英語運用科或General Certificate of Education (Advanced Level) (GCE A Level) English Language科C級或以上成績,會獲接納為等同綜合招聘考試英文運用試卷的二級成績。香港高級程度會考中國語文及文化、中國語言文學或中國語文科C級或以上成績會獲接納為等同綜合招聘考試中文運用試卷的二級成績。持有上述成績的申請人,將不會被安排應考英文運用及/或中文運用試卷。

因應職位要求而報考

香港中學文憑考試英國語文科第4級成績,會獲接納為等同綜合招聘考試英文運用試卷的一級成績。香港中學文憑考試中國語文科第4級成績會獲接納為等同綜合招聘考試中文運用試卷的一級成績。持有上述成績的申請人,可因應有意投考的公務員職位的要求,決定是否需要報考英文運用及/ 或中文運用試卷。

香港高級程度會考英語運用科或GCE A Level English Language科D級成績，會獲接納為等同綜合招聘考試英文運用試卷的一級成績。香港高級程度會考中國語文及文化、中國語言文學或中國語文科D級成績會獲接納為等同綜合招聘考試中文運用試卷的一級成績。持有上述成績的申請人，可因應有意投考的公務員職位的要求，決定是否需要報考英文運用及/或中文運用試卷。

在International English Language Testing System (IELTS)學術模式整體分級取得6.5或以上，並在同一次考試中各項個別分級取得不低於6的人士，在考試成績的兩年有效期內，其IELTS成績可獲接納為等同綜合招聘考試英文運用試卷的二級成績。持有上述成績的申請人，可據此決定是否需要報考英文運用試卷。

綜合招聘考試與公務員招聘

　　一般來說，應徵學位或專業程度公務員職位的人士，需在綜合招聘考試的英文運用及中文運用兩張試卷取得二級或一級成績，以符合有關職位的語文能力要求。個別招聘部門/職系會於招聘廣告中列明有關職位在英文運用及中文運用試卷所需的成績。在英文運用及中文運用試卷取得二級成績的應徵者，會被視為已符合所有學位或專業程度職系的一般語文能力要求。部分學位或專業程度公務員職位要求應徵者除具備英文運用及中文運用試卷的所需成績外，亦須在能力傾向測試中取得及格成績。

　　部分公務員職系（如紀律部隊職系）會按受聘者的學歷給予不同的入職起薪點。未具備所需的綜合招聘考試成績的學位持有人仍可申請這些職位，但不能獲得學位持有人的起薪點。

　　除非有關招聘廣告另有訂明，有意投考學位或專業程度公務員職位的人士，應先取得所需的綜合招聘考試成績。申請人可報考全部、任何一張或任何組合的試卷。申請人應先確定擬投考職位的要求及被接納為等同綜合招聘考試成績的其他考試成績，以決定所需報考的試卷。

綜合招聘考試與公務員職位的招聘程序是分開進行的。有意投考公務員職位的人士，應直接向招聘部門/職系提交職位申請。取得所需的綜合招聘考試成績並不代表考生已完全符合任何學位或專業程度公務員職位的入職要求。招聘部門/職系會核實職位申請人的學歷及/或專業資格，並可能在綜合招聘考試外，另設其他考試/面試。

基本法及香港國安法測試與公務員招聘

由2022年7月1日起刊登的公務員職位招聘，在《基本法及香港國安法》測試取得及格成績是所有公務員職位的入職條件。

《基本法及香港國安法》測試會考核申請人的《基本法》及《香港國安法》知識。無論公務員職位申請人過往曾否參加《基本法》測試或在個別局/部門公務員職位（學位/專業程度職系）招聘過程中安排的《基本法》測試及取得何等成績，都必須在《基本法及香港國安法》測試中取得及格成績，方會獲考慮聘用。

過往曾參加任何《基本法》測試的人士如欲應徵2022年7月1日起刊登的公務員職位招聘，亦須參加《基本法及香港國安法》測試並取得及格成績，方會獲考慮聘用。

公務員職系要求一覽

	職系	入職職級	英文運用	中文運用	能力傾向測試
1.	會計主任	二級會計主任	二級	一級	及格
2.	政務主任	政務主任	二級	二級	及格
3.	農業主任	助理農業主任 / 農業主任	一級	一級	及格
4.	系統分析 / 程序編製主任	二級系統分析 / 程序編製主任	二級	一級	及格
5.	建築師	助理建築師 / 建築師	一級	一級	及格
6.	政府檔案處主任	政府檔案處助理主任	二級	二級	-
7.	評稅主任	助理評稅主任	二級	二級	及格
8.	審計師	審計師	二級	一級	及格
9.	屋宇裝備工程師	助理屋宇裝備工程師 / 屋宇裝備工程師	一級	一級	及格
10.	屋宇測量師	助理屋宇測量師 / 屋宇測量師	一級	一級	及格
11.	製圖師	助理製圖師 / 製圖師	一級	一級	-
12.	化驗師	化驗師	一級	一級	及格
13.	臨床心理學家（衛生署、入境事務處）	臨床心理學家（衛生署、入境事務處）	一級	一級	-
14.	臨床心理學家（懲教署、香港警務處）	臨床心理學家（懲教署、香港警務處）	二級	一級	-
15.	臨床心理學家（社會福利署）	臨床心理學家（社會福利署）	二級	二級	及格
16.	法庭傳譯主任	法庭二級傳譯主任	二級	二級	及格
17.	館長	二級助理館長	二級	一級	-
18.	牙科醫生	牙科醫生	一級	一級	-
19.	營養科主任	營養科主任	一級	一級	-
20.	經濟主任	經濟主任	二級	二級	-
21.	教育主任（懲教署）	助理教育主任（懲教署）	一級	一級	-
22.	教育主任（教育局、社會福利署）	助理教育主任（教育局、社會福利署）	二級	二級	-
23.	教育主任（行政）	助理教育主任（行政）	二級	二級	-
24.	機電工程師（機電工程署）	助理機電工程師 / 機電工程師（機電工程署）	一級	一級	及格
25.	機電工程師（創新科技署）	助理機電工程師 / 機電工程師（創新科技署）	一級	一級	-
26.	電機工程師（水務署）	助理機電工程師 / 機電工程師（水務署）	一級	一級	及格
27.	電子工程師（民航署、機電工程署）	助理電子工程師 / 電子工程師（民航署、機電工程署）	一級	一級	及格
28.	電子工程師（創新科技署）	助理電子工程師 / 電子工程師（創新科技署）	一級	一級	-

	職系	入職職級	英文運用	中文運用	能力傾向測試
29.	工程師	助理工程師 / 工程師	一級	一級	及格
30.	娛樂事務管理主任	娛樂事務管理主任	二級	二級	及格
31.	環境保護主任	助理環境保護主任 / 環境保護主任	二級	二級	及格
32.	產業測量師	助理產業測量師 / 產業測量師	一級	一級	-
33.	審查主任	審查主任	二級	二級	及格
34.	行政主任	二級行政主任	二級	二級	及格
35.	學術主任	學術主任	一級	一級	
36.	漁業主任	助理漁業主任 / 漁業主任	二級	二級	及格
37.	警察福利主任	警察助理福利主任	二級	二級	-
38.	林務主任	助理林務主任 / 林務主任	一級	一級	及格
39.	土力工程師	助理土力工程師 / 土力工程師	一級	一級	及格
40.	政府律師	政府律師	二級	一級	
41.	政府車輛事務經理	政府車輛事務經理	一級	一級	
42.	院務主任	二級院務主任	二級	二級	及格
43.	新聞主任（美術設計）/（攝影）	助理新聞主任（美術設計）/（攝影）	一級	一級	-
44.	新聞主任（一般工作）	助理新聞主任（一般工作）	二級	二級	及格
45.	破產管理主任	二級破產管理主任	二級	二級	及格
46.	督學（學位）	助理督學（學位）	二級	二級	-
47.	知識產權審查主任	二級知識產權審查主任	二級	二級	及格
48.	投資促進主任	投資促進主任	二級	二級	-
49.	勞工事務主任	二級助理勞工事務主任	二級	二級	及格
50.	土地測量師	助理土地測量師 / 土地測量師	一級	一級	-
51.	園境師	助理園境師 / 園境師	一級	一級	及格
52.	法律翻譯主任	法律翻譯主任	二級	二級	
53.	法律援助律師	法律援助律師	二級	二級	及格
54.	圖書館館長	圖書館助理館長	二級	二級	及格
55.	屋宇保養測量師	助理屋宇保養測量師 / 屋宇保養測量師	一級	一級	及格
56.	管理參議主任	二級管理參議主任	二級	二級	及格
57.	文化工作經理	文化工作副經理	二級	二級	及格
58.	機械工程師	助理機械工程師 / 機械工程師	一級	一級	及格
59.	醫生	醫生	一級	一級	-
60.	職業環境衛生師	助理職業環境衛生師 / 職業環境衛生師	二級	二級	及格
61.	法定語文主任	二級法定語文主任	二級	二級	

	職系	入職職級	英文運用	中文運用	能力傾向測試
62.	民航事務主任（民航行政管理）	助理民航事務主任（民航行政管理）/ 民航事務主任（民航行政管理）	二級	一級	及格
63.	防治蟲鼠主任	助理防治蟲鼠主任 / 防治蟲鼠主任	一級	一級	及格
64.	藥劑師	藥劑師	一級	一級	-
65.	物理學家	物理學家	一級	一級	及格
66.	規劃師	助理規劃師 / 規劃師	二級	二級	及格
67.	小學學位教師	助理小學學位教師	二級	二級	-
68.	工料測量師	助理工料測量師 / 工料測量師	一級	一級	及格
69.	規管事務經理	規管事務經理	一級	一級	-
70.	科學主任	科學主任	一級	一級	-
71.	科學主任（醫務）（衛生署）	科學主任（醫務）（衛生署）	一級	一級	-
72.	科學主任（醫務）（食物環境衛生署）	科學主任（醫務）（食物環境衛生署）	一級	一級	及格
73.	管理值班工程師	管理值班工程師	一級	一級	-
74.	船舶安全主任	船舶安全主任	一級	一級	-
75.	即時傳譯主任	即時傳譯主任	二級	二級	-
76.	社會工作主任	助理社會工作主任	二級	二級	及格
77.	律師	律師	二級	一級	-
78.	專責教育主任	二級專責教育主任	二級	二級	-
79.	言語治療主任	言語治療主任	一級	一級	-
80.	統計師	統計師	二級	二級	及格
81.	結構工程師	助理結構工程師 / 結構工程師	一級	一級	及格
82.	電訊工程師（香港警務處）	助理電訊工程師 / 電訊工程師（香港警務處）	一級	一級	-
83.	電訊工程師（通訊事務管理局辦公室）	助理電訊工程師 / 電訊工程師（通訊事務管理局辦公室）	一級	一級	及格
84.	電訊工程師（香港電台）	高級電訊工程師 /助理電訊工程師 / 電訊工程師（香港電台）	一級	一級	-
85.	電訊工程師（消防處）	高級電訊工程師（消防處）	一級	一級	-
86.	城市規劃師	助理城市規劃師 / 城市規劃師	二級	二級	及格
87.	貿易主任	二級助理貿易主任	二級	二級	及格
88.	訓練主任	二級訓練主任	二級	二級	及格
89.	運輸主任	二級運輸主任	二級	二級	及格
90.	庫務會計師	庫務會計師	二級	一級	及格
91.	物業估價測量師	助理物業估價測量師 / 物業估價測量師	一級	一級	及格
92.	水務化驗師	水務化驗師	一級	一級	及格

資料截至：2023 年 8 月

PART II 試題類型練習

A. COMPREHENSION

This section aims to test candidates' ability to comprehend a written text. A prose passage of non-technical background is cited. Candidates are required to exercise skills in deciding on the gist, identifying main points, drawing inferences, distinguishing facts from opinion, interpreting figurative language, etc.

Passage 1

What would it take to persuade you to exercise? A desire to lose weight or improve your body figure? To keep heart disease, cancer or diabetes at bay? To lower your blood pressure or cholesterol? To protect your bones? To live to a healthy old age?

You'd think any of those reasons would be sufficient to get Americans exercising, but scores of studies have shown otherwise. It seems that public health experts, doctors and exercise devotees in the media have been using ineffective tactics to entice *sedentary* people to become, and remain, physically active.

For many decades, people have been bombarded with messages that regular exercise is necessary to lose weight, prevent serious disease and foster healthy aging. And yes, most people say they value these goals. Yet a vast majority of Americans—two thirds of whom are overweight or obese—have thus far failed to swallow the "exercise pills."

Now research by psychologists strongly suggest it's time to stop thinking of future health, weight loss and body image as motivators for exercise. Instead, these experts recommend a strategy marketers use to sell products: portray physical activity as a way to enhance current well-being and happiness.

"We need to make exercise relevant to people's daily lives," Michael L. Segar, a research investigator at the Institute for Research on Women and Gender at the University of Michigan, said in an interview. "Everyone's schedule is packed with nonstop to-do's. We can only fit in what's essential." Dr. Segar is among the experts who believe that people will not commit to exercise if they see its benefits as distant or theoretical.

"It has to be portrayed as a compelling behavior that can benefit us today," she said. "People who say they exercise for its benefits to quality of life exercise more over the course of a year than those who say they value exercise for its health benefits." She has come up with an idea for a public service advertisement to promote exercise for working women with families: A woman is shown walking around the block after dinner with her children. "This is great. I can fit in fitness, spend quality time with my kids, and at the same time teach them how important exercise is."

Based on studies of what motivates people to adopt and sustain physical activity, Dr. Segar is urging that experts stop framing moderate exercise as a medical prescription that requires 150 minutes of aerobic effort each week. Instead, public health officials must begin to address "the emotional hooks that make it essential for people to fit in into their *hectic* life." "Immediate rewards are more motivating than distant ones," she added. "Feeling happy and less stressed is more motivating than not getting heart disease or cancer, maybe, someday in the future."

1. How does the author think of the reasons mentioned in paragraph 1?

   A. They are enough to get more Americans exercise.

   B. They can't break away from the help of media.

   C. They seem incapable of making people exercise.

   D. They are against the will of public health experts.

2. The word "*sedentary*" (paragraph 2) probably means _____.

   A. inactive

   B. busy

   C. obese

   D. unhealthy

3. Which of the following inferences can you draw from paragraph 3?

   A. Americans have received a lot of messages about the important of regular exercise.

   B. The number of spam messages about regular exercise has been on the increase for the past few decades.

   C. Many Americans take pills to reduce weight.

   D. We should stop associating regular exercise with weight lost and healthy aging.

4. According to the passage, a vast majority of Americans _____.

   A. do regular exercise every day

   B. are in danger of serious disease

   C. are too busy to keep healthy

   D. failed to insist on doing exercise

5. It can be inferred from the research by psychologists that _____.

   A. It's too late to think of future health

   B. physical exercise can improve happiness

   C. marketers don't have physical problems

   D. body image is the key motivator for exercise

6. Dr. Segar holds that idea that _____.

   A. women need more physical exercise than men

   B. most people lack basic physical theories

   C. instant benefits attract people to exercise

   D. schedules should involve gender differences

7. The word *"hectic"* (paragraph 7) probably means
_____.

A. Leisure

B. Positive

C. Occupied

D. Hopeful

8. According to the passage, which of the following statements is true?

A. Doing exercise helps people feel happy instead of keeping cancer at bay.

B. Aerobic exercise fails to reduce stress, let alone prevent people from contracting diseases.

C. Immediate rewards should be highlighted when encouraging people to engage in physical activity.

D. People who do exercise are proven to be happier and less stressed.

9. Which of the following would be the best title for this passage?

A. A New Approach to Promoting Exercise

B. How Should We Lose Weight?

C. The True Value of Doing Exercise

D. Promoting Exercise for Working Women

10. What is the tone of the passage?

A. Argumentative

B. Informative

C. Biased

D. Farcical

Answers

1. C   In paragraph 2, the author says that "scores of studies have shown otherwise," indicating that "C" is the correct answer.

2. A   It is mentioned that "the media have been using ineffective tactics to entice sedentary people to become, and remain, physically active." We can infer from this that "sedentary" is the opposite of "active".

3. A   The phrase "bombarded with messages" indicates that people (including Americans) have received plenty of messages.

4. D   In paragraph 3, the author says that "a vast majority of Americans ... failed to swallow the exercise pills." In other words, most Americans do not do exercise regularly.

5. B   Paragraph 4 says that "these experts ... portray physical activity as a way to enhance ... happiness," indicating that "B" is the right answer.

6. C   The line "it has to be portrayed ... that can benefit us today" indicates that "C" is the right answer.

7. C   "Hectic" shares a similar meaning with "busy." "Occupied" is the only word out of the four that has the closest meaning.

8. C   The line "immediate rewards are more motivating than distant ones" clearly shows that "C" is the correct answer.

9. A   The passage focuses mainly on how and why we should start making exercise relevant to people's daily lives. In the words, it talks about a new approach to promoting exercise.

10. B   The author illustrates his arguments mainly through facts, quotes and figures, which makes the passage more informative in nature.

Passage 2

It's a common belief that women take fewer risks than men, and the adolescents always plunge in headlong without considering the consequences. But the reality of who takes risks is actually a bit more complicated, according to a new paper which will be published in the August issue of *Current Directions in Psychology Science*, a journal of the Association for Psychological Science. Adolescents can be as cool-headed as anyone, and in some *realms*, women take more risks than men.

A lot of what psychologists know about risk-taking comes from lab studies where people are asked to choose between a guaranteed amount of money or a gamble for a larger amount. But that kind of decision isn't the same as deciding whether you're going to speed on the way home from work, or go bungee jumping. Research in the last 10 years or so has found that the way people choose to take risks on one domain doesn't necessarily hold in other domains.

"The typical view is that women take less risks than men, that it starts early in childhood, in all cultures, and so on," says Bernd Figner of Columbia University and the University of Amsterdam, who co-wrote the paper with Elke Weber of Columbia University. The truth is more complicated. Men are willing to take more risks in finances. But women take more social risks—a category that includes things like starting a new career in your mid-thirties or speaking your mind about an unpopular issue in a meeting at work.

It seems that this difference is because men and women perceive risks differently. "That difference in perception may be partly because of how familiar they are with different situations," Figner says. "If you have more experience with a risky situation, you may perceive it as less risky." Differences in how boys and girls encounter the world as they're growing up may make them more comfortable with different kinds of risks.

Adolescents are known for risky behavior. But in lab tests, when they're called on to think coolly about a situation, psychological scientists have found that adolescents are just as cautious as adults and children. The difference between the lab and the real world, Figner says, is partly the extent to which they involve emotion. In an experiment where adolescents' emotions got triggered strongly, they looked very different from children and adults and took bigger risks, just as observed in real world settings.

"Ultimately we would like to provide knowledge with our research that people can use to make decisions that are beneficial for them in the long term," Figner says. The goal isn't to avoid risk, but by understanding when and how people decide to take risks, he hopes to help people make risky decisions that they won't regret, either immediately after they have made them, or years later.

1. Which of the following inferences can you draw from paragraph 1?

   A. It is scientifically proven that men take more risks than women.

   B. Adults are more cool-headed than adolescents.

   C. Women actually take more risks than men.

   D. It is a misconception that adolescents always rush to make decisions.

2. The word *"realms"* (paragraph 1) can be best replaced by _____.

   A. domains

   B. cultures

   C. some stages of life

   D. lab tests

3. According to the passage, we can know that _____.

   A. men start to take risks earlier than women

   B. adolescents are the ones who take the most risks

   C. men take more risks in finances than women

   D. adolescents are cool-headed enough as for risk-taking

4. Women tend to take more risks when they are asked to _____.

   A. gamble on money

   B. face unknown dangers

   C. go bungee jumping

   D. express themselves in a meeting at work

5. How will adolescents act in the real world if triggered strongly by the surrounding?

   A. They will still be cool-headed and careful enough.

   B. They will get bold enough to take bigger risks.

   C. They will perceive the potential risks.

   D. They will behave the same as adults and children.

6. Which of the following statements might Figner agree with?

   A. The more risky experiences one has, the less dangerous he will feel.

   B. People can learn skills to avoid taking risks from the paper.

   C. The common belief that women take fewer risks than men is wrong.

   D. There is no apparent difference between the results from the lab and the real world.

7. What conclusion can we draw from the passage?

   A. Adolescents are always blindly imprudent.

   B. Men take more risks than women as they're growing up.

   C. People act differently upon risks in different domains.

   D. People seldom regret their risky decisions.

8. What does Figner want to achieve with his research?

   A. To help people make less risky decisions.

   B. To help people avoid making risky decisions that they will regret.

   C. To understand the reasons why people take risks.

   D. To prove that lab experiments about risk-taking behavior are not always accurate.

9. Which of the following would be the best title for this passage?

   A. Gender Differences in Risk Perception

   B. The Long-term Benefits of Risk Management

   C. Who Takes Risks?

   D. Steps to Avoid Risk

10. In which section of a newspaper are you most likely to find this passage?

    A. Front page

    B. Entertainment

    C. Education

    D. Life

Answers

1. D   The lines "It's a common belief … adolescents always plunge in headlong … consequences" and "adolescents can be as cool-headed as anyone" indicate that "D" is correct.

2. A   "Realms" shares a similar meaning with "domains," "areas," etc.

3. C   In paragraph 3, it is stated that "men are willing to take more risks in finances," indicating that "C" is the correct answer.

4. D   In paragraph 3, it is stated that "women take more social risks … speaking your mind about an unpopular issue in a meeting at work." So "D" is correct.

5. B   Paragraph 4 says that when adolescents' emotions got triggered strongly, they would take "bigger risks." So "B" is the best answer.

6. A   As Figner puts it: "If you have more experience with a risky situation, you may perceive it as less risky." So "A" is correct. "C" is wrong because Figner does not explicitly deny the belief in the passage.

7. C   "C" is correct because the passage mainly talks about how different groups of people differ in risk-taking behavior when they face different situations.

8. B   The final paragraph says that Figner "hopes to help people make risky decisions that they won't regret." So "B" is the correct answer.

9. C   "A" is wrong because the passage does not focus only on gender differences. "B" and "D" are wrong because neither risk management nor steps to avoid risk are mentioned.

10. D   "A" and "B" are ruled out because the passage is neither some breaking news nor a piece about the showbiz. "C" is arguably possible but "D" fits better because the passage is about risk-taking, something that readers can relate to their lives. Remember that when two answers seem correct, always choose the best one for the question asked.

Passage 3

Cloud computing is the hot new buzzword in tech these days. But who knew the killer app for this brave new world would be plain old e-mail? Yet that is exactly what's happening. "E-mail has become the easiest workload for customers to move to the cloud," says Chris Capossela, a senior vice president at Microsoft.

What this shift means, basically, is that instead of buying your own computer servers and paying a team of techies to run your e-mail system, you can instead rent e-mail as a service. Microsoft—or Google or IBM—runs your e-mail system on its servers, taking care of security, software updates, and bugs. It's the first step in a larger shift in which, over the next decade, much of the computing that takes place in corporate data centers will migrate out onto the cloud. As it unfolds, it could create new winners and losers among big tech companies.

Microsoft reckons that the average customer can save 30 percent on e-mail by moving to the cloud. A company with 1,000 employees might spend $2 million a year on e-mail, so the savings are significant. Proponents say cloud-based e-mail is not just cheaper, it's also better. With cloud-based e-mail each user can have huge amounts of storage space—25 gigabytes or more—while a traditional "on-premises" e-mail system might allow users only 100 megabytes.

Rexel, a French distributor of electrical equipment, expects to cut its e-mail costs by one third by moving its employees from a *hodgepodge* of systems to a single cloud-based one, operated by Microsoft, says Olivier Baldassari, the company's chief information officer. So far Rexel has shifted 4,000 of its 28,000 employees to a cloud version of Microsoft Exchange and plans to get everyone moved over by the end of 2015.

Microsoft wasn't the first to offer cloud-based mail to corporate customers: it had to play catch-up with Google. But for companies already using Exchange, it's often easier to move employees to the cloud version of what they already know than to switch them over to a new system. *Serena*, a software company in Redwood City, California, last year went from on-premises Exchange to Google's Gmail but is migrating again, to the cloud-based version of Exchange, mostly because its employees are more familiar with it. "Making the change to Google was gut-wrenching for a lot of people," says Ron Brister, Serena's director of IT. "People just weren't getting used to it."

Microsoft sees the cloud as a competitive weapon, a chance to pull customers away from rivals like IBM, which sells an on-premises messaging system called Notes. "Customers are using the cloud as a way to move to Microsoft," Capossela says, citing new customers like Coca-Cola, and Kraft Foods. But wait—IBM claims it's doing the same thing right back to Microsoft. IBM sells a cloud-based mail solution called LotusLive iNotes and has lured away from Microsoft such customers as Panasonic, which is moving 300,000 employees onto an IBM cloud.

One thing Microsoft, Google, and IBM all agree on: the battle over cloud computing is only just beginning. For old-guard tech companies, it offers the chance to add new customers. But it also brings big risks: during big shifts in technology like this one, industry giants can be swept aside as new leaders emerge. Ten years from now, the tech landscape is certain to look very different.

1. What is the main purpose of paragraph 1?

   A. To state that e-mail is the core value of cloud computing.

   B. To explain why people move e-mail to the cloud.

   C. To introduce Chris Capossela as a senior vice president at Microsoft.

   D. To describe what cloud computing is.

2. As to the e-mail customers, moving to the cloud means _____.

   A. buying their own e-mail servers

   B. doing software updates by themselves

   C. recruiting a team for the running

   D. renting e-mail from the server providers

3. The cloud supporters hold the idea that _____.

A. the average customers has to pay 30% more on the cloud computing

B. there will be no big tech companies as the cloud computing unfolds

C. cloud-based mail is better than the traditional system in its storage

D. most of the cloud computing will happen in corporate data centers

4. Which of the following has the similar meaning with "*hodgepodge*" (paragraph 4)?

A. Shelter

B. Mixture

C. Center

D. Server

5. The example of **Rexel** (paragraph 4) is mentioned to show _____.

A. how cloud-computing is gaining popularity in other parts of the world

B. how cloud-computing can help companies reduce staff size

C. how cloud-computing reduces e-mail costs significantly

D. how cloud-computing facilitates staff mobility

6. According to the passage, Google's Gmail
   _____.

   A. is not a cloud-based version and is difficult to use

   B. is issued earlier than Microsoft's similar service

   C. has fully lost its battle with Microsoft's Exchange

   D. is not popular among people in California

7. Which of the following is true about *Serena* (paragraph 5)?

   A. Its staff is more used to Google's Gmail.

   B. It has decided to move employees from on-premises Exchange to Google's Gmail.

   C. It is returning to on-premises Exchange from Google's Gmail.

   D. Its staff had difficulty in familiarizing themselves with Google's Gmail.

8. Which of the following is helpful for IBM to attract customers?

   A. The on-premises mail solution called Notes.

   B. The cooperation with Microsoft in using the cloud.

   C. The cloud-based mail solution called LotusLive iNotes.

   D. The good remarks from Coca-Cola and Kraft Foods.

9. Which of the following statements might the author agree with?

A. There will hardly be any changes in the tech market over the next decade.

B. Google, and IBM will continue to lead the market.

C. Microsoft might eventually lose out to a new company in the cloud computing market.

D. Cloud computing offers more opportunities than risks.

10. Which of the following would be the best title for this passage?

A. The Secret Weapon of Microsoft

B. The Success of Cloud Computing

C. A New Tech Battleground

D. Reducing Cost with Cloud-based Email

Answers

1. A   A "killer app" is any computer program that is so necessary or desirable that it proves the core value of some larger technology. So "A" is correct.

2. D   Paragraph 2 says that "what this shift means … you can instead rent e-mail as a service." So "D" is the right answer.

3. C   It is indicated in paragraph 3 that supporters say "cloud-based e-mail is … also better" and "cloud-based e-mail each user can have huge amounts of storage space … while a traditional 'on-premises' e-mail system … only 100 megabytes." So "C" is correct.

4. B   From "from a hodgepodge of systems to a single cloud-based one" we can infer that "hodgepodge" should mean the opposite of "single". "B" is the best answer.

5. C   The author mentions Rexel to show that its e-mail costs are to be reduced by one third. So "C" is correct.

6. B   The author says that "Microsoft wasn't the first to offer …" and that "it had to … catch-up with Google." So "B" is the correct answer.

7. D   The author says that Serena is migrating from Google's Gmail to the cloud-based version of Exchange because its employees are more familiar with the latter. In other words, they have problems in familiarizing themselves with Google's Gmail.

8. C   Paragraph 6 says that "IBM sells … called LotusLive iNotes and … lured away from Microsoft such customers …" So "C" is the right answer.

9. C   In the final paragraph, the author says that "industry giants can be swept aside as new leaders emerge." In other words, he believes that it is possible for an industry giant to be eventually losing out to a brand new company in the market.

10. C   "C" is correct because the passage focuses on how different companies compete with each other in the cloud computing market.

Passage 4

Few people doubt the fundamental importance of mothers in childrearing, but what do fathers do? Much of what they contribute is simply the result of being a second adult in the home. Bringing up children is demanding, stressful and exhausting. Two adults can support and make up for each other's deficiencies and build on each other's strengths.

Fathers also bring an array of unique qualities. Some are familiar: protector and role model. Teenage boys without fathers are notoriously prone to trouble. The pathway to adulthood for daughters is somewhat easier, but they must still learn from their fathers, in ways they cannot from their mothers, how to relate to men. They learn from their fathers about heterosexual trust, intimacy and difference. They learn to appreciate their own femininity from the one male who is most special in their lives. Most important, through loving and being loved by their fathers, they learn that they are love-worthy.

Current research gives much deeper—and more surprising—insight into the father's role in childrearing. One significantly *overlooked* dimension of fathering is play. From their children's birth through adolescence, fathers tend to emphasize play more than caretaking. The father's style of play is likely to be both physically stimulating and exciting. With older children it involves more teamwork, requiring competitive testing of physical and mental skills. It frequently resembles a teaching relationship: come on, let me show you how. Mothers play more at the child's level. They seem willing to let the child direct play.

Kids, at least in the early years, seem to prefer to play with daddy. In one study of 2-year-olds who were given a choice, more than two-thirds chose to play with their fathers.

The way fathers play has effects on everything from the management of emotions to intelligence and academic achievement. It is particularly important in promoting self-control. According to one expert, "children who roughhouse with their fathers quickly learn that biting, kicking and other forms of physical violence are not acceptable." They learn when to "shut it down."

At play and in other realms, fathers tend to stress competition, challenge, initiative, risk-taking and independence. Mothers, as caretakers, stress emotional security and personal safety. On the playground fathers often try to get the children to swing ever higher, while mothers are cautious, worrying about an accident.

We know, too, that fathers' involvement seems to be linked to improved verbal and problem-solving skills and higher academic achievement. Several studies found that along with paternal strictness, the amount of time fathers spent reading with them was a strong predictor of their daughters' verbal ability.

For sons the results have been equally striking. Studies *unveiled* a strong relationship between fathers' involvement and the mathematical abilities of their sons. Other studies found a relationship between paternal nurturing and boys' verbal intelligence.

1. The first paragraph points out that one of the advantages of a family with both parents is _____.

   A. husband and wife can share housework

   B. two adults are always better than one

   C. the fundamental importance of mothers can be fully recognized

   D. husband and wife can compensate for each other's shortcomings

2. What is the main purpose of paragraph 2?

   A. To discuss the distinctive roles that fathers play in children's growth.

   B. To give examples to show how fathers fulfill the role of a protector

   C. To downplay the role of mothers in helping daughters appreciate their own femininity.

   D. To explain why teenage boys without fathers are usually prone to trouble

3. The word "*overlooked*" (paragraph 3) can be best replaced by _____.

   A. oversaw

   B. ignored

   C. mistaken

   D. experimented

4. According to paragraph 3, one significant difference between the father's and mother's role in child-rearing is _____ .

   A. the style of play encouraged

   B. the amount of time available

   C. the strength of emotional ties

   D. the emphasis of intellectual development

5. Which of the following statements is true?

   A. Mothers tend to stress personal safety less than fathers.

   B. Boys are more likely than girls to benefit from their fathers' caring.

   C. Girls learn to read more quickly with the help of their fathers.

   D. Fathers tend to encourage creativeness and independence.

6. Studies investigating fathers' involvement in child-rearing show that _____ .

   A. this improves kids' mathematical and verbal abilities

   B. the more time spent with kids, the better they speak

   C. the more strict the fathers are, the cleverer the kids

   D. girls usually do better than boys academically

7. According to the passage, which of the following would a mother be most likely to do to her child?

A. Letting the child play freely and out of sight.

B. Teaching the child through roughhousing that biting and kicking are not at all acceptable.

C. Discouraging the child from playing action sports because of risk of injury.

D. Telling the child why it is important to be cautious in a playground.

8. The word "*unveiled*" (paragraph 8) can be best replaced by _____.

A. uncovered

B. refuted

C. predicted

D. guessed

9. The writer's main point in writing this article is _____.

A. to warn society of increasing social problems

B. to emphasize the father's role in the family

C. to discuss the responsibilities of fathers

D. to show sympathy for one-parent families

10. In which section of a magazine are you most likely to find this passage?

A. Arts and Culture

B. Nature

C. Education

D. Medical Care

Answers

1. D   The phrase "two adults" refers to "husband and wife" while "make up for" and "deficiencies" share similar meanings with "compensate" and "short-comings" respectively. So "D" is correct.

2. A   The sentence "fathers … unique qualities" shows the main idea of paragraph two. And "unique" and "distinctive" are synonyms. So "A" is correct.

3. B   "Ignore" and "overlook" are synonyms.

4. A   From "fathers tend to emphasize play … father's style of play is … both physically stimulating and exciting," we can infer that "A" is the correct answer.

5. D   Paragraph 6 says that "fathers tend to stress … initiative, risk-taking and independence." So "D" is the answer.

6. A   The last two paragraphs say that "fathers' involvement … linked to improved verbal …" and that "relationship between fathers' involvement and the mathematical abilities …" We can therefore infer that "A" is correct.

7. C   Paragraph 6 mentions that mothers "stress … personal safety" and they worry about accidents. We can therefore infer that "C" is the best answer.

8. A   "Unveiled" and "uncovered" share similar meanings.

9. B   In the very beginning, the author asks "what do fathers do?" and then goes on explaining the different roles of fathers with examples. We can therefore infer that "B" is the best answer.

10. C   "C" is the best answer because the passage talks mainly about how the roles of parents matter in terms of children's growth and learning.

Passage 5

Five leading food companies have introduced a labeling scheme for their products in the British market, in an attempt to assuage critics who say they encourage obesity. But consumer groups are unhappy all the same. Is the food industry, like tobacco before it, about to be engulfed by a wave of lawsuits brought on health grounds?

Keeping fit requires a combination of healthy eating and regular exercise. On the second of these at least, the world's food companies can claim to be setting a good example: they have been working up quite a sweat in their attempts to fend off assaults by governments, consumer groups and lawyers who accuse them of *peddling* products that encourage obesity. This week saw the unveiling of another industry initiative: five leading food producers—Danone, Kellogg, Nestlé, Kraft and PepsiCo—introduced a labeling scheme for the British market which will show "guideline daily amounts" for calories, fats, sugar and salt on packaging. The new labels will start to appear on the firms' crisps, chocolate bars, cheese slices and the like over the next few months. A number of other food giants, such as Cadbury Schweppes and Masterfoods, have already started putting guideline labels on their products.

The food companies say doing this will empower consumers, allowing them to make informed decisions about which foods are healthy. But consumer groups have cried foul. They point out that the Food Standards Agency, a government watchdog, is due to recommend a different type of labeling scheme next month: a "traffic light" system using colors to tell consumers whether products have low, medium or high levels of fat, salt and the like. The food firms, they say, have rushed to introduce their own, *fuzzier* guidelines first in a cynical attempt to undermine the government's plan—which they fear might hurt their sales. In consumer tests, the traffic light performed better than rival labeling schemes.

Nevertheless, the food companies argue that the traffic-light system is too simplistic and likely to scare people away from certain products that are fine if consumed in moderation, or in conjunction with plenty of exercise—which most observers, including the medical profession, agree is crucial for anyone wanting to stay in shape. They also point out that they have competitors to worry about—namely the big supermarket chains with their own-label products. Last April, Tesco, the biggest of these, announced that it was rejecting the traffic-light system in favor of a less stark "signposting" approach. Its rivals fear that adopting color-coding could put them at a competitive disadvantage.

1. Why do the five leading food companies decide to label their products?

   A. Because they care about the consumers' health and remind them to notice the nutrition.

   B. Because they are forced to label their food according to the corresponding laws.

   C. Because they want to assuage critics who say they encourage obesity.

   D. Because they are willing to promote their products through activities nice to customers.

2. According to the passage, which of the following statements is true?

   A. The tobacco industry was once overwhelmed by many lawsuits brought on health grounds.

   B. Consumer groups are unhappy because five leading food companies have introduced a labeling scheme for their products in the British market.

   C. Kraft is one of the five leading food companies which has been sued by consumer groups for encouraging obesity.

   D. Five leading food companies in Britain will introduce a labeling scheme that promotes healthy eating and regular exercise.

3. The word *"peddling"* (paragraph 2) can be best replaced by _____.

   A. introducing

   B. recalling

   C. selling

   D. labeling

4. According to consumers, of which of the following features should qualified food inform them?

   A. Having a guideline daily amounts of some harmful ingredients.

   B. Having a notice of the deadline of valid period.

   C. Having guideline labels on their products.

   D. Having a "traffic light" system using colors.

5. The word *"fuzzier"* (paragraph 5) most probably means _____.

   A. more apparent

   B. more blurred

   C. more transparent

   D. more obvious

6. Why does the "traffic light" system receive much opposition?

   A. It does harm to the direct benefit of large food companies.

   B. Customers find it difficult to realize the real content of the food they buy.

   C. "Traffic light" system is sometimes confusing most of the customers.

   D. Food companies regard it as misleading potential customers.

7. Which of the following is true about the Food Standards Agency?

   A. It carried out research that shows the benefits of the "traffic light" system.

   B. It is going to recommend the "traffic light" system some time in the future.

   C. It slightly disapproves of the labeling scheme recently introduced by leading food companies.

   D. It aims to help consumers make informed decisions about food.

8. Which of the topics will most probably be discussed in the following passage?

   A. The drawbacks of "traffic light" system and some vivid examples.

   B. The advantages of "signposting" method to most producers.

   C. The unexpected results after applying for the "traffic light" system.

   D. The similarities between "signposting" and "traffic light" systems.

9. What is the tone of the writer throughout the passage?

   A. Sarcastic

   B. Persuasive

   C. Indifferent

   D. Objective

10. Where are you most likely to find this passage?

   A. In an academic journal.

   B. In an economics textbook.

   C. In a business magazine.

   D. In a talk-show program.

Answers

1. C   Paragraph 1 states clearly that the companies have decided to label their products to "assuage critics who say they encourage obesity." So "C" is the correct answer.

2. A   The question "is the food industry, like tobacco before it, about to be engulfed by a wave of lawsuits brought on health grounds" indicates that the tobacco industry was filled with health-related lawsuits.

3. C   "Peddling" and "selling" share similar meanings.

4. D   Paragraph 3 mentions that "consumer groups … point out …a traffic light system using colors to tell consumers …" So "D" is correct.

5. B   "A", "C" and "D" are wrong because they do not fit in logically.

6. D   Paragraph 4 says that " food companies argue … traffic-light system is too simplistic and likely to scare people away from certain products ..." So "D" is the best answer.

7. B   Paragraph 3 says that "the Food Standards Agency … is due to recommend a … labeling scheme next month." So "B" is the correct answer.

8. B   The author shifts the topic to "signposting" by the end of the passage without revealing too much information about that. It is therefore highly possible that what follows will revolve around this new topic.

9. D   "D" is correct because the author discusses the topic using a third-person point of view.

10. C   Clearly the passage is about a recent incident in the commercial world and is therefore most likely to be seen in a business magazine.

Passage 6

Let's not mince words: ***college can be tough***. According to a 2007 study by the American College Health Association, 43 percent of students reported having felt "so depressed it was difficult to function" at least once in the prior year. Other studies, based on student surveys, suggest that one in five undergraduates reported having an eating disorder, one in six had ***deliberately*** cut or burned himself and one in 10 had considered suicide.

Given those numbers, it's deeply troubling that in 2007 just 8.5 percent of students used their college's counseling services. In other words, students were more likely to consider killing themselves than to seek help. Students feel more afraid to discuss mental-health problems. They think they'll be labeled as the crazy kid who'll shoot up the school.

Counselors say that while they do keep an eye out for students who might pose a risk to others, the overwhelming majority of their patients are no threat to anyone but themselves. Counseling services must look for new ways to reach out to troubled students and let them know that seeking treatment is a strong, smart thing. At Harvard, students can win iPods for attending mental-health screening sessions and are invited to "pajama party" panels, where flannel-clad counselors dispense milk and cookies along with advice about the importance of sleep. "There's still a high level of stigma," says Richard Kadison, head of Harvard's mental-health services. "We're trying to find creative ways of getting the message out."

Many campuses also offer online services allowing students to complete informal diagnostic quizzes away from the prying eyes of their peers. The results are confidential, but can help nudge students toward counseling services. Besides, many colleges encourage parents to pitch in, whether by watching out for warning signs or by coaxing their kids to seek help. *Philadelphia University* now issues students' relatives with a calendar highlighting the toughest times of the year for freshmen, while the University of Minnesota offers online workshops, where parents can learn about conditions such as anxiety and Asperger's syndrome.

Still, students and counselors agree that the most effective outreach programs are those led by students themselves. Semmie Kim, a neuroscience major who founded MIT's chapter of Active Minds in 2007, has held events like a bubble-wrap stomp to help students vent pre-exam stress. She insists that her group's most important role is to provide troubled peers with a sympathetic ear. "We want to make students realize they're not alone," she says. "College will always be tough, but there's no need to suffer in silence."

1. The statement *"college can be tough"* (paragraph 1) is used to introduce _____.

   A. the severity of mental illnesses of college students

   B. the fierce competition of earning scholarship

   C. the feeble relations between teachers and students

   D. the alarming rate of suicide among college students

2. The word *"deliberately"* (paragraph 1) can be best replaced by _____.

   A. unconsciously

   B. accidentally

   C. intentionally

   D. randomly

3. According to paragraph 2, few troubled students turn to counselors for help because they _____.

   A. will be asked to leave school

   B. will receive many screening tests

   C. are afraid to be laughed at by peers

   D. will pay more for counseling services

4. According to the passage, counselors are bothered because they do not know _____.

   A. whether they should tell the truth to their patients directly

   B. how much time they will spend on each screening session

   C. what treatment should be included in their counseling services

   D. how to make counseling services acceptable and available to troubled students

5. Which of the following inferences can you draw from paragraph 3?

   A. A majority of troubled students pose a risk not only to themselves but to their peers.

   B. Counseling services are still considered by many a stigma on campus.

   C. Students at Harvard can win iPods for taking advice about sleep.

   D. Harvard is the first university to promote counseling services though "pajama party" panels.

6. It can be inferred from paragraph 4 that _____.

   A. all universities in America have offered online counseling services

   B. parents play a vital role in solving students' mental-health problems

   C. University of Minnesota took the initiative to give lectures on mental health

   D. Asperger's syndrome is a common developmental disorder that worries many parents in Minnesota

7. The example of *Philadelphia University* (paragraph 4) is mentioned to show _____.

   A. how colleges involve parents in identifying and offering help to troubled students

   B. why it is important to highlight the toughest times of the year for freshmen

   C. what can be done to help students with anxiety

   D. who parents should turn to when their kids need help

8. According to paragraph 5, the most effective mental-health service at college is _____.

   A. the comfort and help from peers

   B. the advice from online workshops

   C. the love and warmth from teachers

   D. the treatment from neuroscience experts

9. According to the passage, Semmie Kim is most probably a _____.

    A. professional counselor

    B. neuroscience expert

    C. college student

    D. relative of a troubled kid

10. Which of the following would be the best title for this passage?

    A. The Depressing Lives of College Students

    B. College Is Tough

    C. Tips to Avoiding Stress in College

    D. Coping with Mental Illnesses on Campus

Answers

1. A The first paragraph says that a considerable number of students suffer from a range of problems such as depression, eating disorder and suicidal thoughts. So "A" is the correct answer.

2. C "Deliberately" and "intentionally" are synonyms.

3. C "C" is the most possible answer because the other three are not at all mentioned in paragraph 2.

4. D Paragraph 3 says that counselors "must look for new ways to reach out to troubled students," indicating that they currently have problems connecting with these students. So "D" is the correct answer.

5. B From the quote ("there's still a high level of stigma") of Richard Kadison, a campus staff member, we can infer that "B" is correct.

6. B "B" is correct because "many colleges encourage parents to pitch in" and "parents can learn about conditions such as anxiety ..."

7. A Philadelphia University is mentioned as an example to elaborate on how "colleges encourage parents to pitch in." So "A" is the best answer.

8. A "A" is correct because "the most effective outreach programs are those led by students themselves."

9. C Semmie Kim is mentioned as an example to elaborate on how outreach programs led by students are effective. We can therefore infer that Semmie is a student.

10. D The whole passage focuses on the different approaches to offering help to mentally troubled students. So "D" is the most possible answer.

Passage 7

Whether mobile phones can cause cancer remains unresolved. But they are also accused by some of causing pain. A growing number of people around the world claim to be "electrosensitive", in other words physically responsive to the electromagnetic fields that surround phones and the other electronic devices that clutter the modern world. Indeed, at least one country, Sweden, has recognized such sensitivity as a disability, and will pay for the dwellings of sufferers to be screened from the world's electronic smog.

The problem is that, time and again, studies of those claiming to be electrosensitive show their ability to determine whether they are being exposed to a real electric field or a sham one is ***no better than chance***. So, unless they are lying about their symptoms, the cause of those symptoms needs to be sought elsewhere.

Michael Landgrebe and Ulrich Frick, of the University of Regensburg, in Germany, think that the "elsewhere" in question is in the brain and, in a paper presented recently to the Royal Society in London, they describe an experiment which, they think, proves their point.

Dr. Landgrebe and Dr. Frick used a body scanner called a functional magnetic-resonance imager to see how people's brains react to two different kinds of stimulus. Thirty participants, half of whom described themselves as electrosensitive, were put in the imager and told that they would undergo a series of trials in which they would be exposed either to an active mobile phone or to a heating device called a thermode, whose temperature would be varied between the trials. The thermode was real. The mobile phone, however, was a dummy.

The type of stimulus, be it the authentic heat source or the sham electromagnetic radiation, was announced before each exposure and the volunteers were asked to rate its unpleasantness on a five-point scale. In the case of heat, the two groups' descriptions of their experiences were comparable. So, too, was their brain activity. However, when it came to the sham-phone exposure, only the electrosensitives described any sensations—which ranged from *prickling* to pain. Moreover, they showed neural activity to match.

This suggests that electrosensitivity, rather than being a response to electromagnetic stimulus, is *akin to* well-known psychosomatic disorders such as some sorts of tinnitus and chronic pain. A psychosomatic disorder is one in which the symptoms are real, but are induced by cognitive functions such as attitudes, beliefs and expectations rather than by direct external stimuli.

The paradoxical point of Dr. Landgrebe's and Dr. Frick's experiment is that mobile phones do indeed inflict real suffering on some unfortunate individuals. It is just that the electromagnetic radiation they emit has nothing whatsoever to do with it.

1. According to the first paragraph, Sweden _____.

   A. has recognized electrosensitivity as a disability

   B. has too many mobile phones

   C. has too many people claiming to be "electrosensitive"

   D. has a serious problem of electronic smog

2. The phrase "*no better than chance*" (paragraph 2) most probably means _____.

   A. quite successful

   B. based on facts

   C. correct only by luck

   D. wrong

3. In the experiment of Dr. Landgrebe and Dr. Frick, _____.

   A. thirty participants described themselves as electrosensitive

   B. the temperature remained the same

   C. the thermode was sham

   D. the mobile phone was sham

4. Which of the following is true about the experiment of Dr. Landgrebe and Dr. Frick?

   A. Only one of the subject groups was exposed to both the thermode and the mobile phone.

   B. Both subject groups had similar responses when exposed to the thermode.

   C. When exposed to the mobile phone, both subject groups gave comparable descriptions of their experiences.

   D. Some participants experienced chronic pain when exposed to the sham electromagnetic radiation.

5. The word "*prickling*" (paragraph 5) can be best replaced by _____.

   A. headache

   B. stinging

   C. coughing

   D. bleeding

6. According to Dr. Landgrebe and Dr. Frick, electro-sensitivity _____.

   A. is not necessarily a response to electromagnetic stimulus

   B. does not have real symptoms

   C. occurs only in some unfortunate individuals

   D. is induced by direct external stimuli

7. The phrase "***akin to***" (paragraph 6) most probably means _____.

   A. different from

   B. caused by

   C. exposed to

   D. similar to

8. According to the passage, the experiment of Dr. Landgrebe and Dr. Frick _____.

   A. has great scientific value

   B. proves mobile phones can cause cancer

   C. shows that the electromagnetic radiation of mobile phones does not inflict real suffering on some people

   D. confirms that those who claim to be electrosensitive are lying about their symptoms

9. In which section of a newspaper are you most likely to find this passage?

   A. Environment

   B. Lifestyle

   C. Economy

   D. Science

10. Which of the following would be the best title for this passage?

A. Sham Radiation Causes Real Pain

B. The Paradox of Mobile Phones

C. The Growing Problem of Electrosensitivity

D. Symptoms of Psychosomatic Disorders

Answers

1. A   "A" is correct because "at least one country, Sweden, has recognized such sensitivity as a disability ..."

2. C   Paragraph 2 says that people who claim to be electrosensitive cannot always tell correctly whether they are being exposed to a real electric field. So "C" is the best answer.

3. D   As opposed to the real thermode, the mobile phone was a dummy. This indicates that "D" is the correct answer.

4. B   "Comparable" and "similar" are synonyms.

5. B   "Prickling" and "stinging" share similar meanings.

6. A   "A" is correct because the passage says that "electrosensitivity, rather than being a response to electromagnetic stimulus, is ..."

7. D   "D" is correct because the author is comparing electrosensitivity to psychosomatic disorders.

8. C   The final paragraph says that "mobile phones do indeed inflict real suffering on some unfortunate individuals." What makes the experiment paradoxical is that it shows the contrary. So "C" is the correct answer.

9. D   "D" is correct because the passage is research-driven and involves scientific experiment.

10. A   "B", "C" and "D" fail to describe the main theme of the passage.

Passage 8

The recession of 2008-09 was remarkable in rich countries for its intensity, the subsequent recovery for its weakness. *The labor market has also broken the rules*, as new research from the OECD, a think-tank of mainly rich countries, shows in its annual Employment Outlook.

Young people always suffer in recessions. Employers stop hiring them; and they often get rid of new recruits because they are easier to *sack*. But in previous episodes, such as the recessions of the 1970s, 1980s and 1990s, older workers were also booted out. This time is different. During the financial crisis in 2008, and since, they have done better than other age groups.

Why have older employees done so well? In some southern European countries they benefit from job protection not afforded to younger workers, but that did not really help them in past recessions. What has changed, says Stefano Scarpetta, head of the OECD's employment directorate, is that firms now bear the full costs of getting rid of older staff. In the past early-retirement schemes provided by governments (in the mistaken belief that these would help young people) made it cheaper to push grey-haired workers out of the door. *These* have largely stopped.

Job losses among older workers have also been more than offset by falls in inactivity, reflecting forces that were already apparent before the crisis. Older workers are healthier than they used to be and work is less physically demanding. They are also more attractive to employers than prior generations. Today's 55- to 64-year-olds are the advance squad of the post-war baby-boomers who benefited from better education than their predecessors.

Older workers now have a sharper incentive to stay in employment because of the impact of the crisis on wealth. In Britain, for example, workers who rely on private pensions have been adversely affected by lower returns on their investments and by poor annuity rates when they convert their savings into regular income.

Many will argue that older workers have done better at the expense of the young. That view is wrongheaded. First, it is a *fallacy* that a job gained for one person is a job lost for another; there is no fixed "lump of labor". And second, as the report shows, young and old people are by and large not substitutes in the workplace. They do different types of work in different types of occupation: younger people gravitate to IT firms, for example, whereas older folk tend to be employed in more traditional industries. There are plenty of things that should be done to help the young jobless, but shunting older workers out of the workplace is not one of them.

1. It can be inferred from paragraph 1 that _____.

   A. the recession of 2008-09 was intense across the globe

   B. the recovery following the recession of 2008-09 was remarkable for its weakness

   C. the recession of 2008-09 did not at all affect poor countries

   D. the OECD is comprised entirely of rich countries

2. The sentences *"The labor market has also broken the rules"* in paragraph 1 implies that in the recession of 2008-09, _____.

   A. most of the older workers were booted out

   B. young people have suffered badly

   C. older workers experienced less impact than younger workers

   D. employers have stopped recruiting young people

3. The word *"sack"* (paragraph 2) can be best replaced by _____.

   A. recruit

   B. train

   C. retain

   D. lay off

4. What can we learn about the early-retirement schemes provided by governments?

    A. They are not afforded to younger workers.
    B. They have made the lives of the old in the recession harder.
    C. Firms now bear the full costs of firing older staff because of the schemes.
    D. They aimed to help young people find and retain their jobs.

5. What does *"these"* (paragraph 3) refer to?

    A. Grey-haired workers
    B. Firms that bear the full costs of getting rid of older staff
    C. Past recessions
    D. Past early-retirement schemes

6. Which of the following is the reason why job losses among older workers have been declining?

    A. Older workers now take initiative in their job hunts.
    B. Much of the work today requires less skill and labor.
    C. Older workers are well-educated and experienced.
    D. Older workers are in better shape than their young counterparts.

7. The word "*fallacy*" (paragraph 6) can be best replaced by _____.

   A. misconception

   B. belief

   C. fact

   D. failure

8. The author rejects the view that older workers have done better at the expense of the young because _____.

   A. older workers are restricted to traditional industries

   B. the young and the old seldom compete for the same type of jobs

   C. the amount of work available to laborers is fixed

   D. older workers are not so aggressive and ambitious as the young

9. The main topic of the passage is about _____.

   A. whether older workers should be blamed for young people's unemployment

   B. the phenomenon that older workers have done well since the financial crisis

   C. how older workers survive the economic downturn and recession

   D. the special impact of the recession of 2008-09 on older workers

10. What is the tone of the writer in paragraph 6?

    A. Ironic
    B. Persuasive
    C. Optimistic
    D. Objective

Answers

1. B "Subsequent" shares a similar meaning with "following".

2. C "B" is wrong because "Young people always suffer in recessions." Paragraph 2 says that the financial crisis in 2008 is different because older workers "have done better than other age groups." So "C" is the correct answer.

3. D "Get rid of" means "remove" and "D" is the only option that shares a similar meaning.

4. C Paragraph 3 talks about the past early-retirement schemes and says that "firms now bear the full costs of getting rid of older staff." So "C" is the best answer.

5. D The author says that "in the past early-retirement schemes provided by governments …" and then goes on saying that "these have largely stopped." We can therefore infer that "D" is correct.

6. A The phrase "falls in inactivity" indicates that older workers are now more active. So "A" is the best answer.

7. A "A" is correct because the author tries to explain why it is wrong to think that "a job gained for one person is a job lost for another."

8. B "B" is correct because the author says that the young and the old "do different types of work in different types of occupation."

9. B The author tells us that the recession of 2008-09 is different because older employees have done better during the crisis. And then he continues to cite facts and quotes to explain why this happens. So "B" is the best answer. The other options are wrong because they fail to reflect the main theme of the passage.

10. B "A" and "C" are easily ruled out. "B" is a better choice over "D" because the author's stance is made explicit especially in the final paragraph when he says "there are plenty of things that should be done … but shunting older workers out of the workplace is not one of them."

B. ERROR IDENTIFICATION

Knowledge on use of the language is tested through identification of language errors which may be lexical, grammatical or stylistic.

Each of the sentence below may contain a language error. Identify the part (underlined and lettered) that contains the error or choose "(E) No error" where the sentence does not contain an error.

1. It is the interaction between people, rather than the events that occur in their lives, that are the main focus of social psychology.

    A. It is

    B. interaction

    C. between

    D. that are

    E. No error

2. Despite much research, there are still certain elements in the life cycle of the insect that is not fully understood.

    A. Despite

    B. in

    C. that is

    D. understood

    E. No error

3. <u>Some</u> snakes have hollow teeth <u>are called</u> fangs <u>that they</u> use to <u>poison</u> their victims.

   A. Some

   B. are called

   C. that they

   D. poison

   E. No error

4. When I consider how <u>talented he is</u> <u>as a vocalist</u>, I cannot help <u>but believing</u> that <u>the public</u> will appreciate his gift.

   A. talented he is

   B. as a vocalist

   C. but believing

   D. the public

   E. No error

5. <u>That</u> the man <u>was saying</u> was so important that I asked everyone <u>to stop</u> talking and <u>listen</u>.

   A. That

   B. was saying

   C. to stop

   D. listen

   E. No error

6. <u>Nearly</u> 75 percent <u>of</u> <u>the</u> land of the Canadian province of British Columbia <u>is</u> covered by forests.

   A. Nearly

   B. of

   C. the

   D. is

   E. No error

7. Those full-time students expected to <u>offer</u> some jobs <u>on</u> campus <u>during</u> the <u>coming</u> winter vacation.

   A. offer

   B. on

   C. during

   D. coming

   E. No error

8. The camera was of <u>so</u> inferior quality <u>that</u> I took it <u>back</u> and asked for a better <u>one</u>.

   A. so

   B. that

   C. back

   D. one

   E. No error

9. Your teacher <u>would have been</u> happy to give you a makeup examination <u>had you gone</u> and <u>explained</u> that your parents <u>were</u> ill at the time.

   A. would have been

   B. had you gone

   C. explained

   D. were

   E. No error

10. <u>At no time</u> in the history of mankind <u>women have had</u> <u>greater opportunities</u> for following careers <u>than</u> they have now.

    A. at no time

    B. women have had

    C. greater opportunities

    D. than

    E. No errors

11. Neither rain nor snow <u>keeps</u> the postman from delivering our letters <u>which</u> we <u>so much</u> look forward to <u>receive</u>.

    A. keeps

    B. which

    C. so much

    D. receive

    E. No error

12. The man regretted <u>having blamed</u> his wife <u>for</u> the mistake, for he later <u>discovered</u> that it was his <u>own</u> fault.

    A. having blamed

    B. for

    C. discovered

    D. own

    E. No error

13. <u>As far as</u> I am concerned, her politics <u>are</u> rather conservative <u>compared</u> with other <u>politicians</u>.

    A. As far as

    B. are

    C. compared

    D. politicians

    E. No error

14. The <u>invigilator</u> asked <u>them</u> who <u>had completed</u> their tests to leave the room as <u>quietly</u> as possible.

    A. invigilator

    B. them

    C. had completed

    D. quietly

    E. No error

15. <u>Never</u> before <u>I have</u> seen anyone <u>who has</u> the skill Ken has when he <u>repairs</u> computers.

    A. Never

    B. I have

    C. who has

    D. repairs

    E. No error

16. This is the <u>longest</u> flight <u>I have ever taken</u>. <u>By the time</u> we get to Korea, we <u>had flown</u> for 10 hours.

    A. longest

    B. I have ever taken

    C. By the time

    D. had flown

    E. No error

17. <u>If</u> anyone does not pick up <u>their</u> dry-cleaning within two months, the management is not <u>obligated</u> to return it <u>back</u>.

    A. If

    B. their

    C. obligated

    D. back

    E. No error

18. There is <u>few</u> evidence that children in language class-rooms learn foreign languages <u>any better</u> <u>than</u> adults in similar <u>classroom situations</u>.

    A. few

    B. any better

    C. than

    D. classroom situations

    E. No error

19. <u>Hardly</u> <u>had he arrived</u> at the office <u>when</u> his phone calls came <u>in rapid succession</u>.

    A. Hardly

    B. had he arrived

    C. when

    D. in rapid succession

    E. No error

20. When I <u>heard</u> the chairman <u>called</u> my name, I walked <u>onto</u> the stage <u>to receive</u> my certificate.

    A. heard

    B. called

    C. onto

    D. to receive

    E. No error

21. <u>No matter</u> poor one <u>may be</u>, one can always find something <u>to be</u> <u>thankful for</u>.

    A. No matter

    B. may be

    C. to be

    D. thankful for

    E. No error

22. I'd rather you <u>would go</u> by train, because I <u>can't bear</u> the idea of <u>your being</u> in an airplane in <u>such</u> a bad weather.

    A. would go

    B. can't bear

    C. your being

    D. such

    E. No error

23. <u>Had it not been</u> your help, we'd <u>never have been</u> able to <u>get over</u> the <u>difficulties</u>.

    A. Had it not been

    B. never have been

    C. get over

    D. difficulties

    E. No error

24. <u>At</u> the next <u>faculty meeting</u>, some of the professors want to discuss <u>about</u> the teaching schedule <u>for</u> the new semester.

    A. At

    B. faculty meeting

    C. about

    D. for

    E. No error

25. I did not mind their <u>coming late</u> to the lecture, but I <u>did</u> <u>object</u> their <u>making so much</u> noise.

    A. coming late

    B. did

    C. object

    D. making so much

    E. No error

26. <u>Standing</u> in the front row <u>was</u> the boy <u>whom</u> every-one thought <u>would be chosen</u> as the winner by the judges.

    A. Standing

    B. was

    C. whom

    D. would be chosen

    E. No error

27. <u>On</u> the ground floor he <u>had</u> a store <u>where</u> he sold <u>canned goods</u>.

   A. On

   B. had

   C. where

   D. canned goods

   E. No error

28. The sun <u>warms</u> the earth, <u>this</u> makes <u>it</u> possible for plants <u>to grow</u>.

   A. warms

   B. this

   C. it

   D. to grow

   E. No error

29. <u>With</u> a view to <u>saving</u> the South China tiger <u>by</u> extinction, a group of local and private organizations <u>initiated</u> a rescue program.

   A. With

   B. saving

   C. by

   D. initiated

   E. No error

30. <u>Founded</u> in 1985 and <u>employed</u> an <u>estimated</u> 36,000 people, the organization has grown <u>to become</u> one of the biggest property management firms in Asia.

   A. Founded

   B. employed

   C. estimated

   D. to become

   E. No error

31. I <u>used to</u> be <u>keen of</u> all <u>scientific</u> subjects but now I <u>would prefer</u> to study art.

   A. used to

   B. keen of

   C. scientific

   D. would prefer

   E. No error

32. It was <u>such a</u> nice day <u>that</u> they <u>decided to</u> have a picnic <u>in the field</u>.

   A. such a

   B. that

   C. decided to

   D. in the field

   E. No error

33. <u>Over</u> the past few years Korean pop music <u>has become</u> <u>increasing</u> popular <u>among youngsters</u> in Hong Kong.

   A. Over

   B. has become

   C. increasing

   D. among youngsters

   E. No error

34. To control quality and <u>making</u> decisions about production are <u>among</u> <u>the many</u> responsibilities of an <u>industrial</u> engineer.

   A. making

   B. among

   C. the many

   D. industrial

   E. No error

35. <u>Unable</u> to see their business <u>as a separate entity</u>, many people fail to make a distinction <u>between</u> their company and <u>them</u>.

   A. Unable

   B. as a separate entity

   C. between

   D. them

   E. No error

36. The fact <u>for which</u> a good teacher has some of the <u>gifts</u> of a good actor does not mean that he or she will <u>indeed</u> be able to <u>act well</u> on stage.

   A. for which

   B. gifts

   C. indeed

   D. act well

   E. No error

37. <u>Neither</u> of the <u>boys</u> who <u>have</u> been studying here <u>know</u> the importance of integrity.

   A. Neither

   B. boys

   C. have

   D. know

   E. No error

38. <u>Despite</u> of the efforts <u>to promote</u> domestic production, the country still relies <u>heavily</u> on <u>imports</u> for many things.

   A. Despite

   B. to promote

   C. heavily

   D. imports

   E. No error

39. It is ironic <u>that</u> James Dean made an advertisement warning teens of how harmful <u>it was</u> to drive <u>fast</u>, yet he himself <u>died from</u> a speeding accident.

    A. that

    B. it was

    C. fast

    D. died from

    E. No error

40. It <u>aims to</u> establish a minimum list of standards that <u>ought to</u> <u>include</u> in all codes of conduct <u>covering</u> labor practices.

    A. aims to

    B. ought to

    C. include

    D. covering

    E. No error

Answers

1. D (that is)

2. C (that are)

3. B (called)

4. C (but believe)

5. A (What)

6. E

7. A (be offered)

8. A (such)

9. E

10. B (have women had)

11. D (receiving)

12. E

13. D (politicians')

14. B (those)

15. B (have I)

16. D (will have flown)

17. D (delete "back")

18. A (little)

19. E

20. B (call)

21. A (However)

22. A (went)

23. E

24. C (delete "about")

25. C (object to)

26. C (who)

27. E

28. B (which)

29. C (from)

30. B (employing)

31. B (keen on)

32. E

33. C (increasingly)

34. A (make)

35. D (themselves)

36. A (that)

37. D (knows)

38. A (In spite)

39. E

40. C (be included)

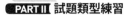

C. SENTENCE COMPLETION

In this section, candidates are required to fill in the blanks with the best options given. The questions focus on grammatical use.

Complete the following sentences by choosing the best answers from the options given.

1. _____ in the desert is mainly due to the limited supply of desert water.

   A. Plants are widely spaced

   B. The spacing of plants is wide

   C. Plants to be spaced widely

   D. The wide spacing of plants

   E. Plants which are spaced widely

2. Over the past few years, intensive design and development effort _____ to the introduction of electronic exchanges.

   A. have been applied

   B. is being applied

   C. has been applied

   D. would have been applied

   E. would be applied

3. Not until the boy turned six _____.

   A. when was his talent discovered

   B. that his talent was discovered

   C. was his talent been discovered

   D. the discovery of his talent

   E. was his talent discovered

4. The result has turned worse than _____ .

   A. would otherwise have been

   B. would be otherwise

   C. has otherwise been

   D. had otherwise been

   E. it otherwise is

5. The knee is _____ most other joints in the body because it cannot twist without injury.

   A. much likely to be damaged than

   B. more likely to be damaged than

   C. likely to be more damaged than

   D. more than likely to be damaged

   E. most likely to be damaged than

6. The bank is reported in the newspaper _____ in broad daylight yesterday.

   A. to be robbed

   B. to have been robbed

   C. robbed

   D. having been robbed

   E. was robbed

7. The business is risky. But _____, we would be rich.

    A. would we succeed

    B. might we succeed

    C. could we succeed

    D. shall we succeed

    E. should we succeed

8. _____ that the pilot couldn't fly through it.

    A. So the storm was severe

    B. So severe was the storm

    C. The storm so severe was

    D. Such was the storm severe

    E. The storm was such severe

9. Prof. Chen's book will show you _____ can be applied in other contexts.

    A. that you have observed

    B. that how you have observed

    C. how that you have observed

    D. how what you have observed

    E. how you have observed that

10. Although pure diamond is colorless and transparent, _____ with other material it may appear in various colors.

    A. but when contaminated

    B. but when contaminating

    C. when contaminated

    D. when contaminating

    E. but while contaminated

11. There seemed little hope that the explorer, _____ in the tropical forest, would find his way through it.

    A. to be deserted

    B. having deserted

    C. to have been deserted

    D. who have deserted

    E. having been deserted

12. _____ their works will give us a much better view of how the two schools of thought differ.

    A. To have reviewed

    B. Having reviewed

    C. Reviewing

    D. Being reviewed

    E. Review

13. _____ to speak when the audience interrupted him.

   A. Hardly had he begun

   B. No sooner had he begun

   C. Not until he began

   D. Scarcely did he begin

   E. Not only did he begin

14. The scientist has made yet another wonderful discovery, _____ of great important to the world.

   A. which I think is

   B. in which I think it is

   C. which I think it is

   D. of which I think it is

   E. it is

15. In no country _____ Britain, it has been said, can one experience four seasons in the course of a single day.

   A. better than

   B. worse than

   C. more than

   D. rather than

   E. other than

16. All the tasks _____ ahead of time, they decided to go on holiday for a week.

    A. been fulfilled

    B. were fulfilled

    C. having been fulfilled

    D. had been fulfilled

    E. fulfilled

17. You can't imagine _____ when they received these nice Christmas presents.

    A. how the children were excited

    B. how excited the children were

    C. how excited were the children

    D. the children were how excited

    E. how the children excited

18. Both teams were in hard training because _____ to lose the game.

    A. either was willing

    B. neither were willing

    C. neither of them was willing

    D. not one of them was willing

    E. neither of them were willing

19. The predictions of _____ able to build a human-oid robot vary by hundreds of years.

    A. when science will be

    B. when shall science be

    C. when is science

    D. when science is

    E. when is science be

20. The millions of calculations involved, had they been done by hand, _____ all practical value by the time they were finished.

    A. had lost

    B. would have lost

    C. would lost

    D. should have lost

    E. lost

21. Many visitors praised the magnificent architecture of the Palace, _____ the Forbidden City.

    A. known to foreigners for

    B. known for foreigners to be

    C. known to foreigners as

    D. known for foreigners as

    E. known as foreigners to be

22. The girl had said little so far, responding only briefly when _____ .

    A. speaking
    B. spoke
    C. speaking to
    D. spoken to
    E. spoke to

23. That the brain, once _____ oxygen, dies has been proven.

    A. depriving of
    B. being deprived
    C. deprived
    D. deprived of
    E. deprives

24. _____ in an atmosphere of simple living was what her parents wished for.

    A. The girl was educated
    B. That the girl had to educate
    C. The girl's being educated
    D. The girl to be educated
    E. The girl had been educated

25. The mother frowned that the dishes had not been washed yet, _____.

   A. and neither had the clothes

   B. and that the clothes had not been washed too

   C. nor the clothes had been

   D. and not the clothes either

   E. and the clothes weren't either

26. There are few areas in the world _____ be grown successfully.

   A. that apricots

   B. apricots that can

   C. where can apricots

   D. can apricots

   E. in which apricots can

27. It is heartening to see millions who had nothing but a record of misery _____ to improve their life.

   A. having had the chance

   B. had the chance

   C. to have the chance

   D. will have the chance

   E. have the chance

28. I'd rather you _____ smoking as soon as possible.

    A. stopping

    B. to stop

    C. had stopped

    D. stopped

    E. stop

29. _____ as a masterpiece, a work of art must transcend the ideals of the period in which it was created.

    A. To rank

    B. The ranking

    C. To be ranked

    D. For being ranked

    E. Be ranked

30. _____ for the timely investment from the general public, our business would not be so successful as it is today.

    A. Were it not

    B. Had it not been

    C. Be it not

    D. Should it not be

    E. Even if

31. Mr. Li, who had been driving all day, suggest _____ at the next town.

    A. to stop

    B. stop

    C. stopping

    D. us to stop

    E. having stopped

32. Neither of the graduates who had applied for a teaching position in the university _____.

    A. has been accepted

    B. have been accepted

    C. will be accepted

    D. were accepted

    E. was accepted

33. Only under certain circumstances _____ to take make-up exams.

    A. students are permitted

    B. permitted are students

    C. will students permit

    D. permit students

    E. are students permitted

34. I think that by the time we find out the cause, we
_____ on this mystery for at least six months.

A. will work

B. will have been working

C. have been working

D. would have been working

E. had been working

35. Either the man or his wife _____.

A. knows why was the money stolen

B. know why the money was stolen

C. know why was the money stolen

D. knows why the money was stolen

E. know why is the money stolen

36. Many technological innovations, such as the telephone, _____ the result of sudden bursts of inspiration were in fact preceded by many inconclusive efforts.

A. whose appearance

B. that appear to be

C. and appear to be

D. which was appeared to be

E. are appearing

37. Working like a telescope, _____ the size of objects at great distances.

 A. which magnifies a telephoto lens

 B. a telephoto lens magnifies

 C. a telephoto lens which magnifies

 D. a telephoto lens to magnify

 E. and magnifying a telephoto lens

38. Certain layers of the atmosphere have special names _____.

 A. which indicated their character properties

 B. whose characteristic properties are indicating

 C. what characterize their indicated properties

 D. that indicate their characteristic properties

 E. characterize their properties

39. _____ to inanimate objects, such as machines, is a form of animism.

 A. When attributing emotion

 B. Attributing emotion

 C. Emotion is attributed

 D. If emotion is attributed

 E. Attribute emotion

40. Perfectly matched pearls, strung into a necklace, _____ a far higher price than the same pearls sold separately.

    A. brings

    B. in order to bring

    C. bringing

    D. bring

    E. their bringing

Answers

1. D   "A" and "B" are ruled out because they cannot function as a subject. "C" and "E" are incorrect as they both wrongly use "plants" as the subject.

2. C   "B," "D" and "E" are incorrect because of the wrong tenses. "Effort" is a singular subject that takes a singular verb, so "C" is the right answer.

3. E   With "not until" at the beginning, we know that it is an inverted sentence and so "A," "B" and "D" are wrong. "C" is grammatically not possible.

4. A   The blank needs a comparative clause that takes no overt subject and has to be in the subjunctive mood. So "A" is correct.

5. B   It is a comparative sentence, so "A" and "E" are ruled out first. "D" does not fit in grammatically.

6. B   The "*be* + reported + to + *infinitive*" structure is used. "B" is chosen over "A" because we need the "to + have + *past participle*" structure to indicate that the incident already happened yesterday.

7. E   The inverted version of "if we should succeed" is "should we succeed," and so "E" is correct.

8. B   The inverted version of the "*subject* + *verb* + so + *adjective* + that" structure is used. Therefore, "the storm was so severe that ..." becomes "so severe was the storm that ..."

9. D   In option D, "what you have observed" works as a noun, as does the whole clause "how what you have observed ... other contexts."

10. C   The "when + *past participle*" structure is used. "When contaminated ..." means the same as "when it is contaminated ..."

11. E   The relative clause "who have been deserted ..." is reduced to "having been deserted." So "E" is correct.

12. C   "C" is correct because "reviewing" works as the subject and it fits logically in the sentence.

13. A   The blank needs an inversion, so "C" is ruled out. "D" and "E" are incorrect because of the wrong tense. "No sooner ..." has to be followed by "than" instead of "when."

14. A   "A" is correct. "I think" is an inserted clause that does not interfere with the structure of the sentence.

15. E   "Other than" means "except for" and is therefore the only option that fits logically in the sentence.

16. C   "C" is correct. "Having" is a present participle used to link two separate sentences: "All the tasks have been fulfilled …" and "They decided to …"

17. B   "D" is ruled out because of the wrong word order. "C" is inverted and cannot function as a noun. "A" and "E" do not fit logically in the sentence.

18. C   "A" does not fit the meaning. "D" is wrong because there are only two teams. "Neither" takes a singular verb, so only "C" is correct.

19. A   "E" is ruled out because of its unintelligibility. The blank needs a noun clause in the future, so "A" is the right answer.

20. B   The "would + have + *past participle*" structure is used to describe a probable result in the past. So "B" is the right answer.

21. C   "known for …" bears a similar meaning with "famous because of …" So "A," "B" and "D" are ruled out.

22. D   The "when + *past participle*" structure is used. "When spoken to …" means "When she was spoken to …"

23. D   The "once + *past participle*" structure is used. "Once deprived of…" means "Once it is deprived of …"

24. C   The main verb is "was", so what comes before it has to be the subject (i.e., noun, noun phrase, noun clause, etc.) This makes "A" and "E" wrong. Both "B" and "E" do not fit the meaning.

25. A   "C", "D" and "E" do not fit the meaning. "B" would be correct had "either" been used instead of "too".

26. E   A relative clause is needed after "in the world", so "B" and "D" are incorrect. "A" and "C" do not fit grammatically.

27. E   The "see + *someone/something + infinitive/gerund*" structure is used. "B," "C" and "D" are therefore incorrect. "A" does not fit the meaning.

28. D   The structure of "would rather + *someone* + *past tense of a verb*" is used to indicate a course of action we would prefer someone else to take. So "A," "B" and "E" are wrong. "C" does not fit the meaning.

29. C   The "to-*infinitive*" structure is used to express the purpose of an action. So "B," "D" and "E" are incorrect. "A" is wrong because "rank" has to be in the passive form in order that the sentence makes sense.

30. B   "Had it not been ..." is the inverted form of "If it had not been ..."

31. C   "Suggest" is followed by a gerund, so "A," "B" and "D" are wrong. "E" is incorrect because it does not fit the meaning.

32. E   "Neither of ..." is a singular subject that takes a singular verb. So "B" and "D" are incorrect. "A" and "C" use the wrong tenses.

33. E   "Only under certain circumstances" has to be followed by an inverted clause, so "A" and "D" are incorrect. "B" does not make sense. "C" is in the active voice which is wrong.

34. B   "B" is correct. The "will + have + been + *V-ing*" structure" is used to describe a continuing or repeating event or action that will have been in progress up to a certain time in the future.

35. D   In the case of "either ... or," the main verb agrees with the subject closest to it. So "B," "C" and "E" are wrong. "A" is incorrect because "why was the money stolen" cannot function as a noun clause.

36. B   "D" and "E" are ruled out because "appear" does not take objects and therefore no passive structure can be formed. A relative clause is needed before "were in fact ..." So "B" the correct answer.

37. B   The blank needs a subject and a main verb, so all options but "B" are ruled out. "Lens" is a singular noun that takes a singular verb.

38. D   The blanks need a relative clause, so "E" is ruled out. "B" and "C" do not make sense. "A" use the wrong tense.

39. B   The blank needs a subject, which makes "B" the only possible answer.

40. D   The blank needs a main verb, so "B," "C" and "E" are wrong. "D" is correct because "Pearls" is a plural noun that takes a plural verb.

D. PARAGRAPH IMPROVE-MENT

In this section, two draft passages are cited. For each passage, questions are set to test candidates' skills in improving the draft. The focus of the questions is on writing skills, not power of understanding.

Passage 1

(1) A lot of people who visit museums, they are unaware that some of the paintings they are admiring are in fact fakes. (2) It's hard to believe that forgers can trick the art experts. (3) But people have been doing it for decades. (4) If a fake is considered to be good enough, the forger may be able to convince people that his work is actually a "lost masterpiece" of a famous artist!

(5) You would think that a trained art historian could take one look at a painting and tell who painted it and when. (6) In 1937 an art expert referred to a newly discovered painting they said was painted in the 17th century as "the masterpiece of Jan Vermeer." (7) But in fact the painting was by a skilled Dutch forger named Hans van Meegeren. (8) Van Meegeren actually bought a real 17th-century painting, and then he cleaned the paint off the canvas, and painted a new picture with a special paint he made himself. (9) The truth was uncovered when van Meegeren was arrested for trying to sell the painting to a foreign buyer. (10) Van Meegeren had to admit that the painting was fake to avoid being sent to jail for smuggling!

1. Which of the following is the best revision of sentence (1), reproduced below?

   A. A lot of people who visit museums, unaware that some of the paintings they are admiring are in fact fakes.

   B. A lot of people who visit museums are unaware of it; some of the paintings they are admiring are in fact fakes.

   C. A lot of people who visit museums unaware, feel that some of the paintings they are admiring are in fact fakes.

   D. A lot of people who visit museums are unaware that some of the paintings they are admiring are in fact fakes.

2. Which of the following is the best revision of sentence (4), reproduced below?

   A. If a fake is good enough, the forger may be able to convince people that his work is actually a "lost masterpiece" of a famous artist!

   B. If it is a fake that is good enough, the forger may be able to convince people that his work is actually a "lost masterpiece" of a famous artist!

   C. If it is considered to be good enough, the forger may be able to convince people that his work is actually a "lost masterpiece" of a famous artist!

   D. Sentence (4) as it is now. No change needed.

3. Which of the following sentences provides the best link between sentences (5) and (6), reproduced below?

    A. Some of the so-called experts may have to go back to school, however.

    B. Many art historians have spent years studying the artistic and technical aspects of painting.

    C. But forgers have techniques for making fresh paint look old that have fooled even the best.

    D. Yet with time the paint will crack and colors will fade.

4. Which of the following is the best revision of sentence (6), reproduced below?

    A. In 1937 an art expert referred to a newly discovered painting said to have been painted in the 17th century as "the masterpiece of Jan Vermeer."

    B. In 1937 an art expert referred to a newly discovered painting they said it was painted in the 17th century as "the masterpiece of Jan Vermeer."

    C. In 1937 an art expert referred to a newly discovered painting said it was painted in the 17th century as "the masterpiece of Jan Vermeer."

    D. Sentence (6) as it is now. No change needed.

5. Which of the following revision best shortens sentence (8), reproduced below?

A. Van Meegeren actually purchased a real 17th-century painting, and then he cleaned the paint, and painted a new picture with a special paint he made himself.

B. Van Meegeren actually purchased a real 17th-century painting, cleaned the paint off the canvas, and painted a new picture with a special paint he made himself.

C. Van Meegeren actually purchased a real 17th-century painting, and then he cleaned the paint off the canvas, then painted with a special paint he made himself.

D. Van Meegeren actually purchased a real 17th-century painting, and then he cleaned the paint off the canvas, and painted a new picture with a special paint.

Answers

1. D   "B" is wrong because it results in a sentence without a verb. "C" is wrong because the pronoun "it" has an unclear referent. "A" is wrong because it drifts away from the original meaning.

2. A   "A" is correct because the painting is properly designated as "a fake" to clarify the subject of the clause, and the unnecessary "considered to be" is omitted.

3. C   "C" is correct because the information about forgers' techniques helps explain why the art historian made the mistake in 1937. "A" is wrong because it fails to acknowledge the skill of forgers in tricking the art experts. "B" is wrong because it does not account for the mistake made by the art historian in 1937.

4. A   "A" is correct because the phrase "said to have been" properly indicates the fact that the painting had been determined by someone to have been painted in the 17th century. "B" is wrong because the pronoun "they" has an unclear referent. "C" is wrong because "it" is redundant. "D" is wrong because it is unclear to whom "they" refers.

5. B   "A" is wrong because the sentence becomes illogical: why would a forger clean the existing paint instead of just washing it off? "C" is wrong because it adds the unnecessary word "then" to what is clearly a chronological sequence.

Passage 2

(1) People nowadays have placed emphasis on the kinds of work that others do, it is wrong. (2) Suppose a man says he is a pilot. (3) Immediately everyone assumes that he is a wonderful person, as if pilots were incapable of doing wrong. (4) However, if you say you're a carpenter or a hawker, some people think that you're not as smart as a doctor or a lawyer. (5) Can't someone just want to do this because they love the work?

(6) Also, who decided that the person who does your taxes is more important than the person who ensures that your house is warm or that your stove works? (7) I know firsthand how frustrating it can be. (8) They think of you only in terms of your job. (9) I used to clean houses in the summer because the money was good; however, all the people whose houses I cleaned seemed to assume that because I was vacuuming their carpets I did not deserve their respect. (10) One woman came into the bathroom while I was scrubbing the tiles. (11) She kept asking me if I had any questions. (12) Did she want me to ask whether to scrub the tiles counterclockwise instead of clockwise? (13) Her attitude made me irritated!

(14) Once I read that the jobs people consider important have changed. (15) My point is, then, that who I want to be is much more important than what I want to be!

1. Which of the following is the best revision of sentence (1), reproduced below?

   A. People nowadays place too much emphasis on the kinds of work that others do.

   B. What kinds of work others do is being placed too much emphasis on by people today.

   C. The wrong emphasis is being placed today on people and what kind of work they do.

   D. Sentence (1) as it is now.　No change needed.

2. Which of the following is the best way to revise and combine sentences (2) and (3), reproduced below?

   A. Suppose a man says he is a pilot, but immediately everyone assumes that he is a wonderful person, as if pilots were incapable of doing wrong.

   B. When a man says he is a pilot, however, immediately everyone assumes that he is a wonderful person, as if pilots were incapable of doing wrong.

   C. Immediately, if they say, for example, he is a pilot, everyone assumes that he is a wonderful person, as if pilots were incapable of doing wrong.

   D. If a man says he is a pilot, for instance, immediately everyone assumes that he is a wonderful person, as if pilots were incapable of doing wrong.

3. Which of the following is the best revision of sentence (5), reproduced below?

   A. Can't someone just want to be a carpenter or a hawker because they love the work?

   B. Can someone just want to do this because they love the work?

   C. Maybe someone just want to be a carpenter or a hawker because they love the work.

   D. Can someone not want to do this although they love the work?

4. Which of the following is the best way to revise and combine sentences (7) and (8), reproduced below?

   A. I know firsthand how frustrating it can be how people think of you only in terms of your job.

   B. I know firsthand how frustrating it can be when they are thinking of one only in terms of your job.

   C. I know firsthand how frustrating it can be when people think of you only in terms of your job.

   D. I know firsthand how frustrating it can be if they think of you only in terms of your job.

5. Which of the following sentences provides the best link between sentences (14) and (15)?

A. Carpenters used to be much more admired than doctors.

B. Doctors, however, are still considered by many today to be among the most trustworthy of professionals.

C. But isn't who I want to be much more important than what I want to be?

D. Even a house cleaner can be important!

Answers

1. A   "A" is correct. The phrase "too much" smoothly embeds a negative judgment within the first clause, thereby making the second clause unnecessary. "B" is wrong because the word order is garbled. "C" is wrong because of wordiness – the sentence does not need both the noun "people" and the pronoun "they."

2. D   "D" is the best answer. In the combined sentence, the dependent clause (appropriately introduced by "if") states a possible condition, and the main clause then describes a likely result. "For instance" indicates that the situation illustrates the statement in sentence (1).

3. A   "A" is correct. In context, the verb "be" is more precise than "do," and the nouns "carpenter" and "hawker" are much more specific than the pronoun "this."

4. C   "A" is wrong because "how" is not an acceptable transition word to link the first clause with the second clause. "B" is wrong because it retains the vague pronoun "they" and introduces another inappropriate pronoun "one."

5. A   "A" is the best answer. It elaborates on sentence (14) and gives a proper example that tells how "the jobs people consider important have changed."

Passage 3

(1) There is one area in school where I have always struggled. (2) Essay writing. (3) When I have an idea that I think will be easy to write, I sit down at the computer. (4) I brainstorm, organize my thoughts, but the words always seem to be jumbled by me. (5) No one else in my family seems to have any of the same problems with essays. (6) Not even my elder sister really understands what I am going through. (7) Everyone else in my family, as far as I can tell, sees writing an essay as a simple process of putting thoughts down on paper.

(8) After school, my tutor does what he can to help me improve my writing. (9) He is skilled at explaining how to assemble thoughts into complete, well organized ideas. (10) He has helped me go over my essays, showed me how to really get to the core of my meaning without meandering and losing focus. (11) He even gives me writing exercises that I can work on between meetings.

(12) Even though I have not mastered the English essay, my skills have improved. (13) I can make my main point understood. (14) I can even, sometimes, write something that can touch and inspire someone. (15) To me, that is a powerful ability, and one that has not come easily. (16) Because I have had to work at it, I understand the true value of writing.

1. Which of the following is the best way to revise and combine sentences (1) and (2), reproduced below?

    A. There is one area in school where I have always struggled and essay writing.

    B. There is one area in school where I have always had to struggle with essay writing.

    C. There is one area in school where I have always had to struggle: essay writing.

    D. There is one area in school where I have always had to struggle, that is essay writing.

2. Which of the following is the best revision of sentence (4), reproduced below?

    A. While I brainstorm and organize my thoughts, I seem to be jumbling the words.

    B. I brainstorm and organize my thoughts, but I always seem to jumble the words.

    C. I brainstormed and organized my thoughts, but my words always seemed to be jumbled.

    D. Sentence (4) as it is now. No change needed.

3. Which of the following is the best revision of sentence (10), reproduced below?

    A. He has helped me go over my essays, showing me how to really get to the core of my meaning without meandering and losing focus.

    B. He has helped me go over my essays, he also showed me how to really get to the core of my meaning without meandering and losing focus.

    C. He helps me go over my essays, and by showing me how to really get to the core of my meaning without meandering and losing focus.

    D. Sentence (10) as it is now. No change needed.

4. Which of the following sentences provides the best link between sentences (12) and (13)?

    A. My thoughts can now be organized into paragraphs.

    B. Sometimes I can also organize my thoughts into paragraphs.

    C. I can organize my thoughts into paragraphs now.

    D. Organizing thoughts into paragraphs is what I can do now.

5. Which of the following is the best revision of sentence (14), reproduced below?

A. And I can write something that can touch and inspire someone.

B. Thinking of this, I can write something that can touch and inspire someone.

C. Sometimes I will also write things that touch and inspire someone.

D. Sentence (14) as it is now. No change needed.

Answers

1. C   "C" is correct. The colon succinctly shows the relationship between the two sentences.

2. B   "A" does not fit logically because the author is not "jumbling the words" while he is brainstorming and organizing his thoughts. "C" unnecessarily switches to passive voice and past tense. "D" is awkward as written because of the use of passive voice.

3. A   "B" is wrong because a comma cannot connect two independent clauses. "C" is wrong because the sentence is not complete. "D" would be acceptable if "and" was added before "showed me how ..."

4. C   "C" is the best answer. It follows the same "I can ..." structure of the following two sentences.

5. D   The idea is well expressed as it is written. It indicates the author's ability to occasionally write something that can "touch and inspire someone," and hints at his pride at being able to do so. So "D" is correct.

Passage 4

(1) No matter what it is we are planning to major in, our math teacher believes we should all study accounting in university. (2) While we may never use these skills in the workplace, he assures us it will come in handy in many areas of our lives. (3) For example, understanding accounting theory will make it easier for us to manage our own personal finances. (4) Even the most basic accounting curriculum will include exercises designed to help strengthen our ability to analyze, summarize, and interpret data. (5) There are also many other courses of study, including economics, which require a thorough grounding in the language of business.

(6) Fundamental accounting principles are not difficult, even students who generally avoid math can master them in a semester. (7) The three basic elements in the accounting process are assets, liabilities, and owner's equity. (8) Assets include anything owned by a company or individual. (9) Owner's equity, which is sometimes referred to as capital or net worth, is the amount by which a company's assets exceed its liabilities. (10) Each of these elements is recorded on a separate financial statement. (11) The financial statements are then used to generate a balance sheet, so named because it demonstrates that the three accounting elements are in balance.

1. Which of the following is the best revision of sentence (1), reproduced below?

   A. No matter what it is we plan to major in; our math teacher believes we should all study accounting in university.
   B. No matter what we plan to major in, our math teacher believes we should all study accounting in university.
   C. Our math teacher believes, no matter what it is we plan to major in, we should all study accounting in university.
   D. Sentence (1) as it is now. No change needed.

2. Which of the following is the best revision of sentence (2), reproduced below?

   A. While we may never use them in the workplace, it is these skills that he assures us will come in handy in many areas of our lives.
   B. While we may never use these skills in the workplace, he assures us, coming in handy in many areas of our lives.
   C. While we may never use these skills in the workplace, he assures us they will come in handy in many areas of our lives.
   D. Sentence (2) as it is now. No change needed.

3. Which of the following is the best revision of sentence (6), reproduced below?

A. Fundamental accounting principles are not difficult; even students who generally avoid math can master them in a semester.

B. Fundamental accounting principles are not difficult, even students whom generally avoid math can master them in a semester.

C. Fundamental accounting principles are not difficult, even students who general avoid math can master it in a semester.

D. Sentence (6) as it is now. No change needed.

4. Which of the following sentences provides the best link between sentences (8) and (9)?

A. Liabilities are then subtracted from total assets.

B. Even money owed to a company is considered an asset.

C. Delivery trucks and office equipment are two examples of assets.

D. A liability is a debt—something owed to someone else.

5. Which of the following is the best revision of sentence (11), reproduced below?

A. The financial statements which are then used to generate a balance sheet, so named because it demonstrates that the three accounting elements are in balance.

B. The financial statements being used to generate a balance sheet, so named because it demonstrates that the three accounting elements are in balance.

C. The financial statements are then used to generate a balance sheet; so named because it demonstrates that the three accounting elements are in balance.

D. Sentence (11) as it is now. No change needed.

Answers

1. B "B" is correct because it eliminates the ambiguous and unnecessary use of the phrase "it is."

2. C Although "workplace" is the noun closest to the pronoun "it," the context tells us that "skills" is the antecedent. "C" is correct because it correctly replaces the singular pronoun "it" with the plural "they."

3. A When two related independent clauses are combined into a single sentence, they should be joined by a semicolon. So "A" is correct.

4. D In this passage, three elements are listed, but only two of them are defined. The missing sentence should contain a brief definition of the missing element "liabilities." So "D" is the best answer.

5. D "A" inserts the pronoun "which," making the sentence into a fragment. "C" misuses the semicolon, since the second clause is not independent.

Passage 5

(1) At almost no cost, young musicians nowadays can put their music online. (2) In the past, it would cost many thousands of dollars to put out a sample tape or CD. (3) Members of a high school rock band rent a recording studio for hundreds of dollars per hour. (4) They might not do their best work in a rush to save money. (5) And they would incur huge expenses to record just a brief performance. (6) It is expensive to send out CDs to agents, radio and TV stations, or any other place where someone might give a listen. (7) Getting a CD into record stores is almost impossible for musicians without a huge following or reputation.

(8) They can record whatever they want, using the lyrics and sounds they like, and post their songs on YouTube, Facebook, and other Web sites. (9) Recording at their own pace, the songs won't be released until the musicians are satisfied that it represents their best work. (10) Some musicians have released full-length albums on the Web. (11) A new award, called "Best Band Web Site," has been added to MTV's annual Video Music Awards.

(12) Using the Internet is no guarantee of success. (13) Not long ago a punk rock group named Thrice-Towed Sloths posted an album on the Web. (14) Within a month over 4,000 listeners had downloaded it.

1. Which of the following is the best revision of sentence (3), reproduced below?

    A. Members of a high school rock band decided to rent a recording studio for hundreds of dollars an hour.

    B. Members of a high school rock band will rent a recording studio for hundreds of dollars per hour.

    C. Members of a high school rock band might rent a recording studio for hundreds of dollars an hour.

    D. Sentence (3) as it is now. No change needed.

2. Which of the following is the best way to revise and combine sentences (4) and (5), reproduced below?

    A. They might not do their best work in a rush to save money and would incur huge expenses to record just a brief performance.

    B. They might not do their best work and rushing to save money, huge expenses would be incurred to record just a brief performance.

    C. They might not do their best work incurring huge expenses in their rush to save money to record just a brief performance.

    D. They might not do their best work rushing to save money, incur huge expenses to record just a brief performance.

3. Which of the following is the best revision of sentence (9), reproduced below?

    A. Recording at their own pace, the songs won't be released or satisfy the musicians until they represent their best.

    B. Recording at their own pace, musicians won't release or be satisfied with any song that represents less than their best work.

    C. Recording at their own pace, musicians know that their songs won't be released and they won't be satisfied until they represent their best work.

    D. Sentence (9) as it is now. No change needed.

4. Which of the following would be the best sentence to insert before sentence (8)?

    A. But things have become so much more convenient nowadays.

    B. No one can deny that it is easier for musicians to publish their works with the help of technology.

    C. Yet an unknown musician can now turn to the Internet to distribute his or her music digitally.

    D. Now, however, tens of thousands of musicians can bypass the usual route to stardom.

5. Which of the following would be the best sentence to insert after sentence (14)?

A. Nowadays, other groups use the Web as their only mode of distribution.

B. The experience of the Thrice-Towed Sloths, however, is a rare exception to the rule.

C. The group is a typical example showing how musicians take advantage of the Internet.

D. The group's music, by the way, is an eclectic mix of buzz-saw drone, twangy guitar, and hardcore drumming.

Answers

1. C   "A" is wrong because the verb "decided" does not fit logically in the text. "B" and "D" introduce incorrect verb tenses.

2. A   "B" is wrong because a comma cannot connect two independent clauses.

3. B   "A" and "D" are ruled out because the subject has to be "musicians" instead of "the songs." "C" is wrong due to the unnecessary addition of the verb "know."

4. D   "A" is wrong because "things" appears confusing. "C" is irrelevant. "D" is the best answer because it helps introduce "YouTube, Facebook ..." in the following sentence.

5. B   "B" is the best answer because it corresponds to the main idea of the paragraph – "using the Internet is no guarantee of success."

Passage 6

(1) Weird and mysterious events happen all the time. (2) We have all had experiences that seemed bizarre, like unexpectedly meeting a long-lost friend when it was the day after dreaming about that friend. (3) Some people refer to these types of events as miracles. (4) Is there any scientific explanation for miracles?

(5) Professor Littlewood defined a miracle as an event with a probability of one in a million. (6) It is unlikely that you would ever see a miracle, isn't it? (7) Well, Littlewood's Law of Miracles states that, for most people, miracles happen roughly once a month. (8) The proof of Littlewood's law is pretty simple. (9) He started by estimating that most people are active—at work or school, interacting with other people—at least 8 hours each day. (10) During this period, we see and hear things almost constantly; Littlewood estimated one new event per second. (11) These events add up fast because 1 per second for 8 hours equals 30,000 per day, or one million in one month.

(12) So if people are exposed to one million events per month, and a miracle is a one-in-a-million event, then we all experience them once a month.

1. Which of the following is the best revision of sentence (2), reproduced below?

A. We have all had experiences that seemed bizarre, like unexpectedly meeting a long-lost friend the day after dreaming about that friend.

B. We have all had experiences that seemed bizarre, like unexpectedly meeting a long-lost friend because it was the day after dreaming about that friend.

C. We have all had experiences that seemed bizarre, like unexpectedly meeting a long-lost friend who we saw the day after dreaming about that friend.

D. Sentence (2) as it is now. No change needed.

2. Which of the following would be the best sentence to insert before sentence (5)?

A. Professor Littlewood taught mathematics at Cambridge University in Britain.

B. Some people insist that miracles cannot be explained by science.

C. It would be useful if we had a scientific law with which to explain miracles.

D. A math professor named Littlewood has tried to explain miracles mathematically.

3. Which of the following is the best revision of sentence (6), reproduced below?

A. At that rate, it is unlikely that you would ever see a miracle, right?

B. In any case, it is unlikely that you would ever see a miracle, isn't it?

C. Unless you are lucky, it is unlikely that you would ever see a miracle, right?

D. Littlewood claimed that it is unlikely that we would ever see a miracle.

4. Which of the following is the best revision of sentence (11), reproduced below?

A. These events add up fast if 1 per second for 8 hours equals 30,000 per day, or one million in one month.

B. These events add up fast: 1 per second for 8 hours equals 30,000 per day, or one million in one month.

C. These events add up to 1 per second for 8 hours because it equals 30,000 per day, or one million in one month.

D. Sentence (11) as it is now. No change needed.

5. Which of the following is the best revision of sentence (12), reproduced below?

A. So if people are exposed to one million events per month, and miracles are one-in-a-million events, then we all experience them once a month.

B. So if people are exposed to one million events per month, and a miracle is a one-in-a-million event, we all experience them once a month.

C. So if people are exposed to one million events per month, and a miracle is a one-in-a-million event, then we all experience a miracle once a month.

D. So if people are exposed to one million events per month, and if a miracle is a one-in-a-million event, then a miracle is experienced by all of us once a month.

Answers

1. A   "A" is correct. It concisely signifies who and when. "C" introduces an illogical cause-effect relationship between the dream and the meeting. "D" includes the redundant phrase "when it was the day after."

2. D   "D" is the best answer because it properly introduces Littlewood and his findings at the beginning of the paragraph and therefore makes a clear transition between sentence (4) and sentence (5).

3. A   "A" is the best choice because the "rate" logically connects this sentence to the previous sentence.

4. B   "B" is correct. The information after the colon properly illustrates the truth of the preceding claim.

5. C   "C" is the best answer because it replaces the vague pronoun "them" with "a miracle."

PART III 模擬考試挑戰

MOCK EXAM PAPER I

MOCK EXAM PAPER I
Time Limit: 45 mins

Reading Comprehension
(10 questions)

Read the following passage and answers questions 1-10. For each question, choose the best answer from the given choices.

It's hard to believe that Dr. Judah Folkman, the pioneering cancer researcher who succumbed to a heart attack on Monday at the age of 74, couldn't ward off death. The man whose mind pulsed with questions, ideas and the arcane details of human biology had survived the most brutal of battles long ago: scientific skepticism. When he first proposed his radical theory of angiogenesis in the 1970s—that cancer tumors grow by recruiting blood vessels for nourishment—he was *derided* by fellow scientists.

Folkman remembered hearing researchers "laughing in the corner" or excusing themselves to go to the bathroom when he got up to speak at scientific meetings. Decades later, in May 1998, a hyperbolic James Watson told the New York Times, "Judah is going to cure cancer in two years." Not so. But angiogenesis *spawned* an entire field of research, led to more than 10 new cancer drugs now on the market (with dozens more in clinical trials), and inspired young researchers to investigate bold new avenues in cancer research.

Moses Judah Folkman didn't seek the limelight. The son of

a rabbi, he spent a lifetime trying to answer the prayers of his patients. He was a healer, a visionary, a compassionate man with a probing intellect and a grandfatherly spirit. During my first interview with him in the midst of the 1998 media glare, Folkman offered me cookies, spent hours poring over the science, then walked me out the front door of Children's Hospital in Boston in his white lab coat to be sure I'd get home safely in a cab.

Some 2,000 newspapers and television crews around the world desperately tried to get his attention. Folkman wasn't interested in being a celebrity though—he refused to be photographed alone for our cover story that week because he didn't want to be singled out for research he insisted was collaborative. He was interested in saving his patients' lives. And it was their lives, not just their medical histories, that mattered. During our interview he shared photographs of each of them as if he were showing family albums; he told me about their hobbies and their dreams. His followers—many of whom call him their hero—believe Folkman should have been awarded the Nobel Prize. Folkman believed he just had to keep asking questions. "You have to think ahead," he once told me. "**Science goes where you imagine it**."

1. According to paragraph 1, Dr. Judah Folkman was outstanding because _____.

    A. his ideas created a stir in blood research

    B. he found better ways to treat heart attack

    C. his theory paved the way for cancer research

    D. his theory unlocked the mystery of human biology

2. The word "***derided***" (paragraph 1) most probably means _____ .

   A. cursed

   B. mocked

   C. criticized

   D. condemned

3. It can be inferred from paragraph 2 that _____ .

   A. more than 10 new cancer drugs were tried in clinics

   B. what James Watson said to the New York Times is not true

   C. other researchers didn't think highly of Folkman's research at the beginning

   D. young researchers can't conduct cancer research independently

4. The word "***spawned***" (paragraph 2) can be best replaced by _____ .

   A. was replaced by

   B. got rid of

   C. was added to

   D. gave rise to

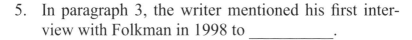

5. In paragraph 3, the writer mentioned his first interview with Folkman in 1998 to _____.

   A. show how intelligent Folkman was

   B. prove that Folkman was not eager to seek the limelight

   C. show that Folkman had a good personality

   D. explain why he would spend hours poring over the science

6. By saying "Science goes where you imagine it," Folkman meant _____.

   A. science would never end

   B. imagination is important in the development of science

   C. scientists should have a spirit of adventure

   D. nothing could be achieved without predictions

7. Which of the following is true about Dr. Judah Folkman?

   A. He was a compassionate grandfather.

   B. He was awarded the Nobel Prize for his achievement.

   C. He refused to be the only one to take credit for collaborative efforts

   D. He was a religious man.

8. In paragraph 4, the writer mentioned Folkman's sharing of photographs to _____.

   A. explain why Folkman's followers call him their hero

   B. emphasize that Folkman did not intend to become famous

   C. show how Folkman and his patients shared similar kind of hobbies and dreams

   D. show that Folkman cared about his patients' lives

9. It can be inferred from the passage that the author is most probably _____.

   A. a journalist

   B. a celebrity

   C. a scientist

   D. a cancer patient

10. Which of the following would be the best title for this passage?

   A. Remembering Judah Folkman

   B. The Death of Judah Folkman

   C. The Scientific Legacy of Judah Folkman

   D. Judah Folkman Dies of Heart Attack

Error Identification

(10 questions)

Each of the sentence below may contain a language error. Identify the part (underlined and lettered) that contains the error or choose "(E) No error" where the sentence does not contain an error.

11. If a hydrogen-filling balloon is brought near a flame, it will explode.

    A. If

    B. hydrogen-filling

    C. near

    D. will explode

    E. No error

12. She cannot tell the difference between true praise and flattering statements make only to gain her favor.

    A. between

    B. flattering

    C. make

    D. to gain

    E. No error

13. One of the <u>wildest</u> and <u>most inaccessible</u> parts of the United States <u>are</u> the Everglades <u>where</u> wildlife is abundant and largely protected.

   A. wildest

   B. most inaccessible

   C. are

   D. where

   E. No error

14. <u>Whenever</u> we <u>hear of</u> a natural disaster, <u>even in</u> a distant part of the world, we feel sympathy for the people <u>to have affected</u>.

   A. Whenever

   B. hear of

   C. even in

   D. to have affected

   E. No error

15. <u>Overseas</u> travel sounds <u>thrilled</u> and sometimes <u>exotic</u> but not everyone can afford <u>it</u>.

   A. Overseas

   B. thrilled

   C. exotic

   D. it

   E. No error

16. This research program <u>is financed by</u> two funds, <u>the largest</u> <u>of</u> which <u>could last</u> for two years.

    A. is financed by
    B. the largest
    C. of
    D. could last
    E. No error

17. <u>As</u> I looked at the carpet, I wished that it <u>could speak,</u> since it <u>must witness</u> many interesting events <u>in</u> the past decade.

    A. As
    B. could speak
    C. must witness
    D. in
    E. No error

18. <u>You'd better</u> walk <u>faster</u> if you want to buy <u>something</u> because there's hardly <u>nothing</u> left.

    A. You'd better
    B. faster
    C. something
    D. nothing
    E. No error

19. I <u>saw that</u> the last update <u>was made</u> three years ago, so I wondered <u>that</u> the webmaster <u>had abandoned</u> the site.

    A. saw that

    B. was made

    C. that

    D. had abandoned

    E. No error

20. A person who uses a personal computer <u>to perform</u> only <u>such tasks as</u> word processing and sending e-mail <u>need not buy</u> the <u>most advanced</u> model available on the market.

    A. to perform

    B. such tasks as

    C. need not buy

    D. most advanced

    E. No error

MOCK EXAM PAPER I

Sentence Completion
(10 questions)

Complete the following sentences by choosing the best answers from the options given.

21. When the famous violinist was small, she _____ for at least six hours a day.

    A. was used to practice

    B. was practiced

    C. was used to be practicing

    D. was used to practicing

    E. used to practicing

22. I spoke to the child in a soft voice _____ him.

    A. not to frighten

    B. so as not to frighten

    C. in order to not frighten

    D. for not frightening

    E. in order that not to frighten

23. _____, the chairperson made the motion to adjourn.

   A. There being no other items for discussion

   B. There were no other items for discussion

   C. There had been no other items to be discussed

   D. There be no other items to be discussed

   E. That there were no other discussion items

24. Raccoons, _____ their food in water, do not actually wash it.

   A. they may dip

   B. they dip

   C. although they may dip

   D. they may although dip

   E. though dip

25. _____ as president is commonly accepted as the main reason for the scandals which ruined his administration.

   A. That the man was indecisive

   B. The man whose indecisiveness

   C. It was that the man was indecisive

   D. The man who was indecisive

   E. Because the man was indecisive

26. Not until the existence of positively charged electrons was discovered in 1933 _____ the theory of negative kinetic energy.

    A. was it possible to confirm

    B. possibly was confirmed

    C. it was possible to confirm

    D. it possibly confirmed

    E. would it have been possible

27. Total color blindness, _____, is the result of a defect in the retina.

    A. that a rare condition

    B. that is a rare condition

    C. a rare condition that

    D. is a rare condition

    E. a rare condition

28. _____, the project will have to be called off.

    A. All things be considered

    B. We have considered all things

    C. Considering all things

    D. All things considered

    E. All things have been considered

29. Until then, his family _____ from him for over a year.

    A. hasn't heard

    B. hadn't heard

    C. didn't hear

    D. wouldn't have heard

    E. wouldn't hear

30. It was not until last night that _____.

    A. did the detective find out the truth

    B. had the detective found out the truth

    C. the detective found out the truth

    D. has the detective found out the truth

    E. was the detective found out the truth

Paragraph Improvement
(10 questions)

The sentences below are parts of the early draft of two passages, some parts of which may have to be rewritten. Read the passages and choose the best answer to the question.

Passage A

(1) Over the past two decades, organic food—food grown without artificial fertilizers or pesticides—has become hugely popular. (2) In 2006 organic food sales in the United States alone reached 17 billion. (3) This figure is quite surprising when you consider that organic food is often much more expensive than conventionally grown food. (4) Organic fruits and vegetables can cost as much as 40 percent more than conventionally grown produce. (5) Many people believe that the extra cost is justified as organic grown food is more healthful than conventionally grown food.

(6) Many proponents of organic food argue that artificial fertilizers and pesticides are harmful to human health and that people should therefore consume only food that has been grown without them. (7) While this is surely a valid point, consumers can reduce their exposure to at least some of these chemicals by peeling or

thoroughly washing produce. (8) They also argue that organically grown produce is more nutritious. (9) Some studies do, in fact, show that organic produce may initially be more nutritious than conventional produce, but other research suggests that storing and shipping may negate any initial nutritional advantage. (10) An organic pear shipped 2,000 miles and then left to sit on a grocery store shelf for two weeks may well have fewer vitamins than does a freshly picked conventionally grown pear.

(11) However, one argument for organic food is undeniable. (12) It is that organic food is better for the environment. (13) The reason for this is that farms that grow organic food use less energy and leave soils in better condition than on conventional farms.

31. Which of the following is the best revision of sentence (4), reproduced below?

A. By comparison, organic fruits and vegetables can cost as much as 40 percent more than conventionally grown produce.

B. But organic fruits and vegetables can cost as much as 40 percent more than conventionally grown produce.

C. Indeed, organic fruits and vegetables can cost as much as 40 percent more than conventionally grown produce.

D. This explains why organic fruits and vegetables can cost as much as 40 percent more than conventionally grown produce.

32. Which of the following would be the best sentence to insert after sentence (5)?

 A. But conventionally grown food also has its benefits.

 B. Some scientists, though, believe that organic food is not necessarily more healthful.

 C. This belief, however, is not totally supported by the evidence.

 D. Is there really proof that supports what they believe?

33. Which of the following is the best revision of sentence (8), reproduced below?

 A. Proponents also argue that organically grown produce is more nutritious.

 B. They have also argued that organically grown produce is more nutritious.

 C. On the contrary, they also argue that organically grown produce is nutritious.

 D. They also argue that organically grown produce is more nutritious by comparison.

34. Which of the following is the best way to revise and combine sentences (12) and (13), reproduced below?

A. Organic food is better for the environment, by organic farms producing food that uses less energy and leaves soils in better condition than conventional farms.

B. Compared to conventional farms, which used more energy and left soils in worse condition, the food grown on organic farms was better for the environment.

C. Being that organic food is grown on farms that use less energy and leave soils in better condition than conventional farms, so it is better for the environment.

D. Organic food is better for the environment because organic farms use less energy than conventional ones and leave soils in better condition.

35. Which of the following would be the best sentence to insert after sentence (13)?

   A. Nonetheless, supermarkets are devoting more and more space to the display of organic foodstuffs.

   B. And, as the cost of energy derived from conventional fuel rises, alternative sources will become economically viable.

   C. Ultimately, whether organic food's lower environmental impact continues to justify its higher cost will be decided by the consumer.

   D. Still, the issue will not be resolved until the government clearly indicates what is "organic' and what is not.

Passage B

(1) Play in young animals is an appealing and pretty mysterious behavior. (2) But unlike many other behaviors, play seems to be biologically purposeless and even disadvantageous. (3) They do not achieve an obvious, life-serving goal, as they do in other behaviors such as finding food, mating, repelling intruders, and resting. (4) In fact, animals at play seem to expend a lot of energy for no useful purpose and risk hurting themselves, attracting predators, or otherwise decreasing their chances of survival. (5) There is the obvious explanation that play is inherently enjoyable, offering the intrinsic reward of pleasure. (6) But surely play must have some additional benefits that increase animals' survival chances and thus outweigh the risks incurred and the energy expended.

(7) Researchers believe that play may have evolved at least in part to enhance the ability of animals to adapt to novel situations. (8) Through play, animals gain knowledge of the properties of objects, sharpen their motor skills, and recognize and manipulate characteristics of their environment. (9) Picture a young dolphin blowing air bubbles while underwater, and then chasing them in order to catch them in his mouth before they can reach the surface and vanish. (10) He is not content to repeal this amusing process endlessly. (11) So he will move closer and closer to the water's surface, forcing himself to work harder each time to catch the bubbles before they disappear. (12) Such behavior demonstrates creativity and the desire for increasingly challenging puzzles. (13)

Thus it is consistent with the notion that play facilitates the development and maintenance of flexible problem-solving skills.

36. Which of the following is the best revision of sentence (3), reproduced below?

   A. When they play, animals do not achieve an obvious, life-serving goal, as they do in other behaviors such as finding food, mating, repelling intruders, and resting.

   B. They are not achieving an obvious, life-serving goal, as they do in other behaviors such as finding food, mating, repelling intruders, and resting.

   C. In playing they do not achieve an obvious, life-serving purpose, as they do in other behaviors such as finding food, mating, repelling intruders, and resting.

   D. Sentence (3) as it is now. No change needed.

37. Which of the following would be the best sentence to insert between sentences (4) and (5)?

   A. Why, then, do young animals devote so much time to play?

   B. Is play truly innate, or can it be taught?

   C. On the other hand, what do humans gain from this?

   D. When did animal play first attract the interest of scientists?

38. Which of the following is the best revision of sentence (7), reproduced below?

A. However, researchers believe that play may have evolved at least in part to enhance the ability of animals to adapt to novel situations.

B. Indeed, researchers believe that play may have evolved at least in part to enhance the ability of animals to adapt to novel situations.

C. Researchers believe that this behavior may have evolved in part to enhance the ability of animals to adapt to novel situations.

D. In fact, some researchers tend to believe that play may have evolved at least in part to enhance the ability of animals to adapt to new situations.

39. Which of the following is the best way to revise and combine sentences (10) and (11), reproduced below?

A. He is not content to repeal this amusing process endlessly, therefore he will move closer and closer to the water's surface, forcing himself to work harder each time to catch the bubbles before they disappear.

B. The dolphin, not being content to repeat this amusing process endlessly, moving closer and closer to the water's surface, forcing himself to work harder each time to catch the bubbles before they disappear.

C. Not content to repeal this amusing process endlessly, the dolphin moves closer and closer to the water's surface, forcing himself to work harder each time to catch the bubbles before they disappear.

D. Repealing this amusing process endlessly does not content the dolphin, so that moving closer and closer to the water's surface, forcing himself to work harder each time to catch the bubbles before they disappear.

40. Which of the following is the best revision of sentence (12), reproduced below?

A. The example demonstrates creativity and the desire for increasingly challenging puzzles.

B. In this way it demonstrates creativity and the desire for increasingly challenging puzzles.

C. The dolphin then demonstrates creativity and the desire for increasingly challenging puzzles.

D. Sentence (12) as it is now. No change needed.

Answers

1. C "C" is correct because the passage says that Dr. Judah Folkman was a "pioneering cancer researcher."

2. B Paragraph 2 says that "Folkman remembered hearing researchers laughing … excusing themselves … at scientific meeting …"

3. C Paragraph 2 describes how Folkman's research was disregarded at first but was later on welcomed as inspiration by young researchers. So "C" is correct.

4. D "Spawn" has a similar meaning with "give rise to."

5. C Paragraph 3 aims to describe how Folkman stayed behind "the limelight" and remained kind. So "C" is correct.

6. B "Goes" corresponds to "development."

7. C The passage says that Folkman "didn't want to be singled out for research he insisted was collaborative." So "C" is correct.

8. D The sharing is mentioned to tell how Folkman's patients' lives mattered.

9. A "A" is the most possible answer because the author once interviewed Folkman in 1998.

10. A "B" and "D" are wrong because the passage does not focus on his death. "A" is chosen over "C" because the latter only deals with a portion of the passage.

11. B (hydrogen-filled)

12. C (made)

13. C (is)

14. D (affected)

15. B (thrilling)

16. B (the larger)

17. C (must have witnessed)

18. D (anything)

19. C (if)

20. E

21. D "D" is correct. The "*be* + used + *V-ing*" structure is used to describe the state of being accustomed to or familiar with something.

22. B "B" is correct. "So as not to" means "for the purpose of avoiding/preventing" and has to be followed by an infinitive.

23. A The blank needs a nominative absolute. So "A" is correct. "B" and "C" are independent clauses. "E" is wrong because it is a noun clause.

24. C "C" is inserted within "Raccoons … wash it" as a subordinate clause.

25. A The blank needs a subject. "A" is a noun clause that can function as a subject, so "A" is correct. "C" is wrong because "it was" is redundant. "E" introduces an illogical cause and effect relationship.

26. A The blank needs an inverted clause, so "B," "C" and "D" are ruled out.

27. E "E" is correct. "A rare condition" is the reduced version of the relative clause "which is a rare condition."

28. D "D" is correct. The "*noun* + *past participle*" structure is used here to begin the sentence.

29. B "Then" signifies a time in the past. The blank therefore needs a verb in the past perfect tense, making "B" the correct answer.

30. C The blank needs the normal "*subject* + *verb*" structure, making C the only possible answer.

31. C "C" is correct. "Indeed" indicates that the author is elaborating on the idea mentioned in sentence (3).

32. C "C" provides the best transition to the second paragraph that talks about how organic food is not always more healthful.

33. A "B" is ruled out because the use of the present perfect tense appears unnecessary and confusing. "C" is wrong because "on the contrary" does not fit logically in the text. "A" is chosen over "D" because "by comparison" is redundant.

34. D "A" is grammatically not possible. "B" is incorrect because it uses the wrong tense. "C" is ruled out because "being that" is unnecessary and confusing.

35. C "C" provides the best conclusion for the passage. "A" is irrelevant. "B" and "D" shift the focus out from what is being discussed.

36. A The pronoun "they" brings confusion. "A" helps connect the pronoun with "animals."

37. A "A" provides the best transition that smoothly switches the focus from the downsides of play to its positive value.

38. B "B" is correct – "indeed" serves to affirm the idea that is brought forward from the previous paragraph. "A" is ruled out because of the wrong connective "however." "C" is not the best choice because "this behavior" is vague.

39. C "C" states clearly the cause-effect relationship.

40. D "C" is ruled out because "then" introduces an unnecessary and confusing sequence of actions. "B" is incorrect because the dolphin does not purposely "demonstrates creativity and ..."

MOCK EXAM PAPER II
Time Limit: 45 mins

Reading Comprehension
(10 questions)

Read the following passage and answers questions 1-10. For each question, choose the best answer from the given choices.

Face it, movie fans: the DVD is destined to be *dead as a doornail*.

Only a few *Blockbuster* stores are still open. Netflix's CEO says, "We expect DVD subscribers to decline steadily every quarter, forever." The latest laptops don't even come with DVD slots. So where are film *enthusiasts* supposed to rent their flicks? Online, of course.

Streaming movies offers instant gratification: no waiting, no driving—plus great portability: you can watch on gadgets too small for a DVD drive, like phones, tablets and super-thin laptops. There are still some downsides to streaming movies—you need a fast Internet connection, for example, and beware the limited-data plan—but overall, this should be a delightful development.

Hollywood movie studios should benefit, too. The easier it is to rent a movie, the more people will do it. And the more folks rent, the more money the studios make.

Well, apparently, none of that has occurred to the movie industry. It seems intent on leaving money on the table.

For all of the apparent convenience of renting a movie via the Web, there are a surprising number of drawbacks. For example, when you rent the digital version, you often have only 24 hours to finish watching it, which makes no sense. Do these companies really expect us to rent the same movie again tomorrow night if we can't finish it tonight? In the DVD days, a typical rental period was three days. Why should online rentals be any different?

When you rent online, you don't get any of the DVD extras—deleted scenes, alternative endings, subtitles—even though you're paying as much as you would have paid to rent a DVD.

New movies aren't available online until months after they are finished in the theaters. Worse, some movies never become available. *Star Wars, Jurassic Park, A Beautiful Mind, Bridget Jones's Diary, Saving Private Ryan*, and so on, are not available to rent from the major online distributors.

And if you don't make your product available legally, guess what? The people will get it illegally. Of the 10 most pirated movies of 2011, guess how many of them are available to rent online? Zero. That's right: Hollywood is actually encouraging the very practice they claim to be fighting (with new laws, for example).

Yes, times are changing. Yes, uncertainty is scary. But Hollywood has case studies to learn from. The music industry and the television industry used to fight the Internet the same way—with brute force: copy protection, complexity, legal challenges.

Eventually all of them found roads to recoup some of their lost profit not by fighting the Internet but by working with it. The music industry dropped copy protection and made almost every song available for about $1 each. The TV industry made its shows available for free at sites such as Hulu, paid for by ads.

1. The phrase *"dead as a doornail"* (paragraph 1) most probably means _____.

    A. unmovable

B. disappearing into history

    C. probably dead

    D. unstable

2. According to the passage, *Blockbuster* (paragraph 2) is most probably _____.

    A. a website that rents movies online

    B. a manufacturer that produces laptops

    C. a chain of stores that rents DVDs

    D. a chain of movie theatres

3. The word *"enthusiasts"* (paragraph 2) can be best re-placed by _____.

    A. fans

    B. makers

    C. studios

    D. stores

4. Which of the following statements might the author agree with?

   A. Streaming movies work best on small gadgets such as phones, tablets and ultra-thin laptops.

   B. The benefits of streaming movies outweigh its drawbacks.

   C. It takes a little time for streaming movies to offer satisfaction.

   D. Most streaming movies are delightfully developed.

5. What is the main purpose of paragraph 5?

   A. To show that the movie industry has decided to take the opportunity to make more money.

   B. To show that it is now easier to rent movies on-line.

   C. To show that the movie industry seems to be giving up the advantages that the Internet offers.

   D. To show that the movie industry has left money only the table for everyone to take.

6. According to the passage, which of the following is not true?

A. You cannot get the DVD extras when renting a movie online because you pay less.

B. You have to finish watching the movie within a day when renting the digital version online.

C. Some movies are still unavailable to be rented online even after they are finished in the theatres.

D. New movies can be seen only months after their release.

7. Which of the following statements might the author agree with?

A. The movie industry might eventually work with the Internet.

B. Some of the industry's lost profit will be regained by continuing the fight.

C. The movies will be made available for free at certain websites.

D. Copy protection will be dropped and every movie will be made available for about $1 each.

8. It can be inferred from the passage that _____.

    A. the movie industry should stick to brutal fighting

    B. copy protection will prevent movies from being pi-
    rated

    C. movies should be made available online for free

    D. online renting is a good way to recoup lost profit

9. Which of the following would be the best title for this
    passage?

    A. The End of DVDs

    B. Time to Rent Movies Online

    C. The Benefits of Streaming Movies

    D. Times are Changing

10. What is the tone of the passage?

    A. Formal

    B. Humorous

    C. Neutral

    D. Casual

Error Identification
(10 questions)

Each of the sentence below may contain a language error. Identify the part (underlined and lettered) that contains the error or choose "(E) No error" where the sentence does not contain an error.

11. All of the plants now <u>raised</u> on farms <u>have been developed</u> from plants <u>once grew</u> <u>wild</u>.

    A. raised

    B. have been developed

    C. once grew

    D. wild

    E. No error

12. <u>Breaking</u> up water <u>into</u> hydrogen and oxygen <u>are</u> a good example of <u>what is known</u> as a chemical change.

    A. Breaking

    B. into

    C. are

    D. what is known

    E. No error

13. <u>With</u> production <u>having gone up</u> steadily, the factory <u>needs</u> an <u>ever-increasing</u> supply of new materials.

A. With

B. having gone up

C. needs

D. ever-increasing

E. No error

14. The amount of <u>pressure</u> <u>which</u> the materials are <u>subject</u> to <u>affect</u> the quality of the products.

A. pressure

B. which

C. subject

D. affect

E. No error

15. <u>The more</u> people <u>praise</u> him, the more <u>modestly</u> he <u>becomes</u>.

A. The more

B. praise

C. modestly

D. becomes

E. No error

16. Some psychologists believe that <u>those who</u> are encouraged to be <u>independence</u>, responsible and competent in childhood are <u>more likely than</u> others to become <u>motivated to achieve</u>.

    A. those who

    B. independence

    C. more likely than

    D. motivated to achieve

    E. No error

17. Saturn is the <u>second largest</u> planet <u>after</u> Jupiter, with a diameter <u>nearly ten times</u> <u>those</u> of Earth.

    A. second largest

    B. after

    C. nearly ten times

    D. those

    E. No error

18. <u>The biggest</u> single hobby in America, the one <u>that</u> Americans <u>spend</u> most time, energy and money <u>is</u> gardening.

    A. The biggest

    B. that

    C. spend

    D. is

    E. No error

19. In the fourteenth century, it <u>was discovered</u> that glass <u>coated</u> with silver nitrate <u>would turn</u> yellow when <u>fired</u> in an oven.

    A. was discovered

    B. coated

    C. would turn

    D. fired

    E. No error

20. A conductor uses <u>signals and gestures</u> to let the musicians <u>to know</u> when <u>to play</u> various parts of a <u>composition</u>.

    A. signals and gestures

    B. to know

    C. to play

    D. composition

    E. No error

Sentence Completion
(10 questions)

Complete the following sentences by choosing the best answers from the options given.

21. Tornadoes, powerful, destructive wind storms, occur most often in the spring when hot winds _____ over flat land encounter heavy, cold air.

   A. which to rise

   B. that rising

   C. are rising

   D. rising

   E. that rises

22. Intended to display the work of twentieth-century artists, _____ in 1929.

   A. the opening of the Museum of Modern Art

   B. so the Museum of Modern Art opened

   C. why the Museum of Modern Art opened

   D. and the Museum of Modern Art opened

   E. the Museum of Modern Art opened

23. I don't remember _____ the chance to try this method.

    A. having been given

    B. to have been given

    C. having given

    D. to have given

    E. give

24. Only recently _____ possible to separate the components of fragrant substances and to determine their chemical composition.

    A. it becomes

    B. having become

    C. has it become

    D. which becomes

    E. was it become

25. Smoking is so harmful to personal health that it kills _____ people each year than automobile accidents.

    A. seven more times

    B. seven times more

    C. over seven times

    D. seven times

    E. more seven times

26. The individual TV viewer invariably senses that he or she is _____ an anonymous, statistically insignificant part of a huge and diverse audience.

A. everything except

B. anything but

C. no less than

D. rather than

E. nothing more than

27. Hydrogen is the fundamental element of the universe _____ it provides the building blocks from which the other elements are produced.

A. since that

B. in that

C. so that

D. owing to that

E. provided that

28. The prophet claims to be able to predict _____ with the use of a crystal ball.

A. the results of the presidential election will be

B. the result will be of the presidential election

C. what result will be of the presidential election

D. what the result of the presidential election will be

E. what will the result of the presidential election be

29. Electrical resistance is a common property of all materials, _____.

    A. differ only in degree

    B. only in degree it differs

    C. differing only in degree

    D. it only differs in degree

    E. and differ only in degree

30. The number of colleges and universities _____ since 1980.

    A. has been on the rise

    B. have been increasing

    C. was increased

    D. are increasing

    E. had increased

Paragraph Improvement
(10 questions)

The sentences below are parts of the early draft of two passages, some parts of which may have to be rewritten. Read the passages and choose the best answer to the question.

Passage A

(1) Many reasons are used to justify the cruel practice of keeping animals penned up in zoos. (2) For one thing, parents bring their children to gawk at the caged creatures. (3) Then, thinking they are being kind to the poor creatures, they drop quarters into food dispensers and toss a few pellets to the monkeys or elephants. (4) There must be better reasons, imprisoning wild animals is simply barbaric.

(5) Some people argue that a zoo is educational by allowing visitors to see what animals look like. (6) If someone is so dumb that they don't know what a zebra looks like, they should look it up online. (7) But humans have no right to pull animals from their natural environment and to seal their fate forever behind a set of cold metal bars. (8) Animals need to run free and live, but by putting them in zoos we are disrupting and disturbing nature.

(9) Then there is the issue of sanitary conditions for animals at the zoo. (10) When the animals have been at the zoo for a while

they adopt a particular lifestyle. (11) They lounge around all day, and they are fed at a particular time. (12) They get used to that. (13) That means that they would never again be able to be placed back in their natural environment. (14) They would never survive. (15) And if they reproduce while in captivity, the offspring are born into an artificial lifestyle.

31. Which of the following is the best revision of sentence (4), reproduced below?

   A. I disagree because there must be better reasons, imprisoning wild animals is simply barbaric.

   B. There has got to be better reasons, imprisoning wild animals is terribly barbaric.

   C. There must be better reasons because imprisoning wild animals is simply barbaric.

   D. There must be better reasons, to imprison wild animals is simply barbaric.

32. Which of the following is the best revision of sentence (6), reproduced below?

   A. They claim that viewing a live animal is much more informative than looking at its picture.

   B. But if someone is so dumb that they don't know what a zebra looks like, they should look it up online.

   C. Reading about animals online rather than studying them firsthand.

   D. Doesn't everyone know what a zebra looks like, even little children?

33. Which of the following is the best revision of sentence (9), reproduced below?

    A. Living conditions for animals in the zoo are ordinarily harsh and cruel.

    B. Living in the zoo, conditions of animals affect them permanently.

    C. Captivity alters the basic nature of animals.

    D. No one favors zoos that deliberately try to change the lifestyle of animals in captivity.

34. Which of the following is the best way to revise and combine sentences (12), (13) and (14), reproduced below?

    A. Growing accustomed to that, placing them back in their native habitat and being unable to survive on their own.

    B. They, having gotten used to being fed regularly, in their natural environment would never survive.

    C. Being unable to survive back in their natural environment, the animals have grown accustomed to regular feedings.

    D. Having grown used to regular feedings, the animals would be unable to survive back in their native environment.

35. Which of the following would be the best sentence to insert after sentence (15)?

A. If, as someone say, zoos serve a useful purpose, then why not put humans on display too?

B. After a few generations the animals become totally different from their wild and free ancestors, and visitors to the zoo see animals hardly resembling those living in their natural habitat.

C. The whole idea of a zoo is inhumane.

D. So zoos should be shut down to save animals from extinction.

Passage B

(1) A significant problem in our city is garbage. (2) Our landfills are full. (3) It seems that we must either find new sites for landfills or employ other methods of disposal, like incineration. (4) Unfortunately, there are drawbacks to every solution that they think of. (5) Polluted runoff water often results from landfills. (6) With incineration of trash, you get air pollution. (7) People are criticized for not wanting to live near a polluting waste disposal facility, but really, can you blame them?

(8) Recycling can be an effective solution, owners of apartment complexes and other businesses complain that recycling adds to their expenses. (9) Local governments enjoy the benefits of taxes collected from business and industry. (10) They tend to shy away from pressuring such heavy contributors to recycle.

(11) Perhaps those of us who are concerned should encourage debate about what other levels of government can do to solve the problems of waste disposal. (12) We should make a particular effort to cut down on the manufacture and use of things that will not decompose quickly. (13) Certainly we should press individuals, industries, and all levels of government to take responsible action while we can still see green grass and trees between the mountains of waste.

36. Which of the following is the best revision of sentence (4), reproduced below?

A. Unfortunately, there are drawbacks to every solution that was thought of.

B. Unfortunately, there are drawbacks to every solution that has been proposed.

C. Unfortunately, there are drawbacks to every solution to which there are proposals.

D. Unfortunately, there are drawbacks to every solution that they have previously come up with.

37. Which of the following is the best way to revise and combine sentences (5) and (6), reproduced below?

A. Landfills often produce polluted runoff water, and trash incineration creates air pollution.

B. With landfills, polluted runoff water will result, and whereas with incineration of trash, you get air pollution.

C. While on the one hand are landfills and polluted runoff water, on the other hand you have air pollution in the case of incineration of trash.

D. Landfills and incineration that produce water and air pollution.

38. Which of the following is the best revision of sentence (8), reproduced below?

A. Although recycling can be an effective solution, mostly owners of apartment complexes and other businesses are complaining that it adds to their expenses.

B. Recycling can be an effective solution, however owners of apartment complexes and other businesses complain that recycling adds to their expenses.

C. Even though recycling can be an effective solution, owners of apartment complexes and other businesses complain that recycling adds to their expenses.

D. Recycling can be an effective solution, and owners of apartment complexes and other businesses complain that it adds to their expenses.

39. Which of the following is the best way to revise and combine sentences (9) and (10), reproduced below?

A. However, local governments enjoy the benefits of taxes collected from business and industry, they tend to shy away from pressuring such heavy contributors to recycle.

B. In addition to enjoying the benefits of taxes collected from business and industry, local governments tend to shy away from pressuring business and industry into recycling.

C. Because local governments enjoy the benefits of taxes collected from business and industry, they tend to shy away from pressuring such heavy contributors to recycle.

D. Local governments, enjoying the benefits of taxes collected from business and industry, they tend to shy away from pressure to recycle.

40. Which of the following is the best revision of sentence (11), reproduced below?

A. Perhaps those concerned ones of us should encourage debate about what other levels of government can do to solve the problems of waste disposal.

B. Perhaps we, being among those who are concerned, should encourage debate about what other levels of government can do to solve the problems of waste disposal.

C. Perhaps those of us being concerned should encourage debate about what other levels of government can do to solve the problems of waste disposal.

D. Sentence (11) as it is now. No change needed.

Answers

1. B "B" is correct. The passage deals with the decreasing popularity of DVDs.

2. C "C" is correct. Paragraph 2 continues to talk about how DVDs are losing its popularity and it clearly states that Blockbusters are "stores."

3. A In can be inferred from the passage that "film enthusiasts" probably means "people who like films." So "A" is the best answer.

4. B "B" is correct. The author says that "but overall … a delightful development."

5. C "C" is correct. Paragraph 5 mentions the drawbacks, so "A" and "B" are ruled out first. "D" is irrelevant.

6. A "A" is the correct answer because "you pay less" is not a reason why people "cannot get the DVD extras when renting a movie online."

7. A "A" is the best answer. The final paragraph says that "eventually … not by fighting the Internet but by working with it."

8. D "D" is correct. The final paragraph says that lost profit can be recouped by working with the Internet.

9. B "A" and "C" are ruled out because both only deal with a portion of the passage.

10. D "B" is ruled out first. "C" is not right because the author does share his own opinions throughout the passage. "D" is chosen over "A" because of the conversational language used.

11. C (that once grew)

12. C (is)

13. B (going up)

14. D (affects)

15. C (modest)

16. B (independent)

17. D (that)

18. B (on which)

19. E

20. B (know)

21. D   "D" is the best answer. "B" and "E" would be correct if "rising" and "rises" are placed by "rise."

22. E   The blank needs an independent clause, so only "E" is correct.

23. A   The "remember + V-ing" structure is used to describe a past action that is remembered. "C" is incorrect because it wrongly uses the active voice.

24. C   The blank needs an inverted clause, so "A," "B" and "D" are ruled out. "E" is incorrect because it wrongly uses the passive voice and the past tense.

25. B   "A" and "E" are incorrect because of the wrong word order.

26. E   "Nothing more than" bears the meaning of "the same as," "merely," etc.

27. B   "In that" is used to show a reason.

28. D   The blank needs a noun clause. So "D" is the answer.

29. C   The blank needs a present participle phrase, making "C" the only possible answer.

30. A   "Number" is a singular subject that takes a singular verb, so "B" and "D" are ruled out. "C" and "E" introduce incorrect verb tenses.

31. C   "C" is the best answer because it eliminates the comma splice, retains the basic meaning of the sentence, and subordinates one idea to another.

32. A   "A" is correct because it develops the point stated in sentence (5). "B" is written in a hostile and inappropriate tone. "C" lacks a main verb, making it only a sentence fragment. "D" is irrelevant.

33. C   "C" introduces the main idea of the paragraph. "A" contains an idea not discussed in the paragraph. "B" contains a dangling modifier – the phrase "living in the zoo" should modify "animals" instead of "conditions."

34. D   "D" accurately and economically conveys the ideas of the original sentences. "A" lacks a main verb, making it only a sentence fragment. "B" is grammatically acceptable but stylistically awkward because the subject "they" is too far removed from the verb "would … survive."

35. B   "B" provides the best conclusion for the main idea of passage. "A" and "D" introduce an unexpected shift in topic. "C" is inconclusive.

36. B   "B" is correct. It avoids the error of the original by providing a passive verb phrase "has been proposed," and gets rid of the vague pronoun "they."

37. A　"A" successfully combines the two sentences using the same grammatical structure in both clauses and linking them with the word "and." "B" is wrong because it involves improper coordination: the two clauses are improperly linked with both "and" and "whereas." "C" is wordy and fails to make clear the link between landfills and polluted runoff water.

38. C　"C" is the best answer – the main clause completes the sentence by describing a negative reaction and needs no connecting word to link it with the introductory dependent clause. "A" is ruled out because the adverb "mostly" is unnecessary and alters the meaning of the original.

39. C　"C" is correct – the subordinating conjunction "because" clearly establishes a cause-effect relationship between the benefits described in the introductory clause and the tendency described in the main clause.

40. D　"D" is correct – the dependent clause "who are concerned" is an appropriate idiom to follow the pronoun "us," and the entire phrase uses no unnecessary words.

PART IV 考生常見疑問

1. 什麼人符合申請資格？

   - 持有大學學位；

   - 現正就讀學士學位課程最後一年；或

   - 持有符合申請學位或專業程度公務員職位所需的專業資格。

2. 「綜合招聘考試」(CRE) 跟「聯合招聘考試」(JRE)有何分別？

   在CRE中英文運用考試中取得「二級」成績後，可投考JRE，考試為AO、EO及勞工事務主任、貿易主任四職系的招聘而設。

3. CRE成績何時公佈？

   考試邀請信會於考前12天以電郵通知，成績會在試後1個月內郵寄到考生地址。

4. 報考CRE的費用是多少？

   不設收費。

5. 如果我在今年綜合招聘考試中不及格，我會否被禁止再次應考未來的綜合招聘考試？

   否。你可以在適當的申請期內，報考未來的綜合招聘考試。有關詳情，請瀏覽公務員事務局網頁（www.csb.gov.hk）。如需查詢有關綜合招聘考試事宜，你可聯絡公務員考試組（電話：（852）2537 6429 或電郵csbcseu@csb.gov.hk。）

6. 我在以前曾經應考綜合招聘考試。請問該試的成績是否仍然有效？

   於2006年12月及以後考獲的綜合招聘考試中文運用及英文運用試卷的二級及一級成績和能力傾向測試的及格成績永久有效。所有在2006年12月以前的綜合招聘考試成績已經無效。

7. 我將會應考International English Language Testing System（IELTS）。其結果會否被確認為等同所需要的綜合招聘考試成績？

   在IELTS學術模式整體分級取得6.5或以上，並在同一次考試中各項個別分級取得不低於6的成績的人士，在IELTS考試成績的兩年有效期內，其成績可獲接納為等同綜合招聘考試英文運用試卷的二級成績。IELTS考試成績必須在職位申請期內任何一日仍然有效。換言之，在2015/16年度的政務主任招聘中，任何由2013年9月19日至2015年10月2日期間所獲得，而又符合上述條件的IELTS成績，將被接納為等同所需綜合招聘考試的成績。

8. 如果我從政務主任、二級行政主任、二級助理貿易主任、二級管理參議主任及二級運輸主任職位中，申請了多於一個職位，我是否需要參加多次筆試？

否。如果你申請了多於一個職位，只要你符合所申請職位的入職條件，你只需要應考同一個招聘考試，即聯合招聘考試。

9. 請問申請多個不同職位的申請人，會否同時被邀請參加不同職位的面試？

如果申請人申請了多於一個職位，他／她有可能同時被邀請參加所申請職位的面試，這視乎他／她是否符合不同職位的入職條件和遴選準則。

10. 有沒有聘請本地／非本地申請人的「配額」？

沒有。正在海外就讀或居住的申請人，也是按照與本地申請人同一套的標準，以評核他們在面試時的表現。

看得喜 放不低

創出喜閱新思維

書名	投考公務員 題解EASY PASS 英文運用（第六版）
ISBN	978-988-76628-9-1
定價	HK$138
出版日期	2023年9月
作者	Ray Leung
責任編輯	Mark Sir、吳淑貞
版面設計	梁文俊
出版	文化會社有限公司
電郵	editor@culturecross.com
網址	www.culturecross.com
發行	聯合新零售（香港）有限公司
	地址：香港鰂魚涌英皇道1065號東達中心1304-06室
	電話：（852）2963 5300
	傳真：（852）2565 0919

網上購買 請登入以下網址：

一本 My Book One　　　香港書城 Hong Kong Book City
（www.mybookone.com.hk）　（www.hkbookcity.com）